Vitaly Gupal

MATHEMATICS AND THE MYSTERIOUS GENETIC CODE

Monograph
Second edition

Academus Publishing, Inc.
2026

Academus Publishing, Inc.
1999 S. Bascom Avenue, Suite 700 Campbell CA 95008
Website: www.academuspublishing.com
E-mail: info@academuspublishing.com

Vitaly Gupal is a mathematician and military systems engineer. Trained in both the engineering and command disciplines of the Soviet military-technical academies, he spent ten years in acceptance testing and quality assurance for the instrumentation of major spacecraft and satellite programs — including deep-space and communications systems — before completing his service at the central office of the Ministry of Defense. His research applies probability theory and mathematical statistics to the study of complex systems, and most recently to the structure of the genetic code.

Translated (from Russian) by artificial intelligence tools. The publisher has carried out technical editing. However, the text may contain translation inaccuracies, typos and artifacts. We would be grateful for feedback to the publisher.

UDC 519.1:575(075.4)

BISAC:
SCI029000 SCIENCE / Life Sciences / Genetics & Genomics
MAT029050 MATHEMATICS / Probability & Statistics / Bayesian Analysis
SCI102000 SCIENCE / Bioinformatics

ISBN 978-1-4946-0030-3 (PDF)
ISBN 978-1-4946-0031-0 (ePub)
DOI: 10.31519/1-4946-0030-7

This monograph addresses current issues in modern genetics and questions regarding the analysis and recognition of the genetic code. Much remains unknown and unexplored in the field of genetics, particularly regarding the genetic coding system. The book presents a comparative analysis of recognition methods. It is noted that the Bayesian approach forms the basis of inductive inference procedures, as it is optimal for the entire class of recognition problems and can be widely used to solve problems in various fields of science, including biology and genetics. An example of a Bayesian recognition procedure is provided for predicting protein secondary structure based on Boolean functions.

For researchers, students of biological disciplines, and mathematicians, as well as for a wide range of readers interested in problems of statistical machine learning theory, bioinformatics, and genetics.

CONTENTS

PREFACE

The book recommended for readers is devoted to mathematical methods used in the field of genetics. The aim of the book is to draw people's attention to this science, considering that the general population in Russia remains largely genetically illiterate. Not coincidentally, Academician Konstantin Georgievich Skryabin once stated: "Whoever possesses genetic information will possess the world." [1]

Genetics continues to be a science that holds many mysteries within it; in particular, its genetic code remains highly mysterious.

It is becoming increasingly clear that genetics, as a branch of biology, cannot do without the precision of a science like mathematics.

The scientific and practical directions of research in the field of genetics were established at the end of the 19th century and the beginning of the 20th century. In 1953, D. Watson and F. Crick published their discovery, which initiated a whole new era of research in the science of life. This discovery was the identification of the double-helical structure of DNA.

Practical medical genetics began to develop rapidly around the world and in Russia relatively recently, in the late 1980s to early 1990s, with the launch of the global international project "Human Genome". During this period, there was a tremendous breakthrough in science that had a significant impact on medicine. However, in Russia, far from all the possibilities offered by modern science are being utilized [2].

In 2013, it marked the 10th anniversary of the announcement of the completion of the sequencing of the human genome.

The result of the "Human Genome" project was the development of new, more advanced research methods in which the majority of the processes were fully automated. The participation of a large number of experts from the United States, China, France, Germany, the United Kingdom, and Japan in this project meant that deciphering the nucleotide sequence of DNA, sequencing, and fully mapping all of human genes could help solve fundamental problems in genetics. However, funding for these efforts was reduced in Russia, and Russia did not actually participate in the sequencing activities. Our country's genomic research program was completely restructured and focused on a new area: bioinformatics, which uses mathematical methods to understand and make sense of all the data that has been obtained. The primary goal of the "Human Genome" project was to study the structure of the genome of Homo sapiens; it also aimed to study the genomes of other organisms used as models for biological experiments. In addition to sequencing and mapping, the project's objectives included:

1) Create a bank and a database containing full-length, cloned DNA sequences.
2) Research on the functions of non-coding sequences.
3) Study of the spatiotemporal parameters of gene expression.
4) Study of mutagenesis;
5) Study of gene expression at the product level.
6) The development of technologies that accelerate and reduce the cost of analyzing [3].

In 2004, researchers from the International Human Genome Sequencing Consortium announced the number of genes in the human genome to be between

20,000 and 25,000. Previously, estimates ranged from 3,000 to 40,000; at the beginning of the project, this number had even been as high as 2 million.

The completion of the "Human Genome Project" contributed significantly to the advancement of molecular biology and genetics as a whole. The scientific methodology and instrumentation used in these studies were improved. The project also spurred the development of new areas within genetics, such as genomics and proteomics. From the very beginning of the genomic research efforts, the participating countries agreed that all the data generated would be made available to the public, regardless of the country's contribution or state affiliation. Today, we have access to extensive databases containing billions of nucleotide pairs from the human genome as well as the genomes of other organisms.

In 2003, the "decoding" of the "Human Genome" was announced. However, the term "decoded" should be considered incomplete, as science has not yet uncovered all the mysteries of the genetic code. To truly "decode" the genome means to understand the meaning of the genetic information. In this context, "decoding" refers to determining which specific nucleotides make up genes in what order, and thus which proteins (with what amino acid sequences) are encoded by those genes. So far, only the coding portion of the genome has been elucidated (and not even in its entirety), while the non-coding portions remain a mystery or nearly a mystery. There is a significant gap between the achievements in structural analysis and the functional understanding of the genome. Scientists realize that this is merely the first step, which has been termed the "structural" phase. Consequently, the second and more crucial step—the "functional" phase—must follow, involving the study of the functions of genes.

How much time it will take to truly decode the human genome remains unknown. The origin of 20 amino acids is still unexplained. There are also a number of other questions for which virtually no one can provide an answer at present.

- Why is the genetic code universal?
- Why is the alphabet of the genetic language composed of four letters?
- Why does the genetic code have a triplet structure?
- Why are there precisely these specific sets and combinations of amino acids, with these particular sizes, and not others?
- Why are there two types of amino acids in proteins – polar and non-polar?
- Why is the code endowed with precisely such a system of degeneracy?
- Why is the genetic code non-overlapping? In other words, why does the cellular translational machinery, which reads the genetic information, use a discrete unit equal to three instead of one?
- Is the emergence of these dialects (the mitochondrial version) accidental or systematic?

Trying to find an appropriate approach to this problem is an urgent need for scientists.

Researchers still have a lot of work to do in describing the genome: identifying all the genes, determining their functions, understanding the states in which the genes exist, and detecting harmful mutations. It is necessary to thoroughly understand how and when genes are activated, as well as how the instructions contained within them determine the properties of the proteins they produce. To achieve this level of description, we need greater accuracy in identifying gene sequences. Currently, the error rate when directly reading a sequence is approximately 0.05–2%, depending on various factors. Thanks to the high speed of analysis, this sequence can be read multiple times, thereby reducing the chances of errors. In the end, the sequencing accuracy can reach one error per 100,000 base pairs—this is comparable to the accuracy achieved by traditional methods. However, when dealing with a genome as large as the human genome (which contains 3 billion base pairs), such an error rate means that approximately 30,000 errors will occur simply due to incorrect readings [4].

The biggest problem that scientists have encountered is the vast, unknown regions of DNA. These are sections of DNA that do not contain genes and therefore do not encode any proteins. It was unclear what purpose they served. Some geneticists referred to these regions as "junk"; others disagreed, arguing that evolution actually gets rid of this "junk" rather than accumulating it. One theory is that the presence of large amounts of non-coding DNA has helped stabilize the genome in terms of mutations—by reducing the frequency of mutations occurring in functional genes. Another hypothesis suggests that non-coding DNA is, at least partially, involved in the formation (production) of different species.

However, the greater the achievements researchers make, the more difficult it becomes for them to obtain new results, and the more questions arise that need to be answered.

Not surprisingly, Francis Collins, the leader of the "Human Genome Project" who replaced James Watson in April 1993, described the completion of the DNA sequencing as merely "the end of the beginning."

However, the importance of the "structural" phase in the study of the human genome should not be underestimated. Nearly all of the goals that the project set for researchers have been achieved.

Two scientific approaches were used in the decoding of the human genome, both of which are based on the analysis of nucleotide base sequences. Each of these approaches has its own advantages and disadvantages; they complement each other and serve as evidence of the reliability of the results obtained.

Computer programs have been developed to analyze the data, as it is practically impossible to interpret the data on its own without such programs. Another milestone has been reached in the effort to complete the "Human Genome Project." In 2007, American and British scientists were awarded the world's highest scientific award for their invention of a method that allows for making changes to the genes of living cells.

Just a few years ago, scientists were facing difficulties in extracting the desired DNA sequences from tissue samples. The process of identifying relevant DNA regions was accelerated by a pattern that had previously gone unnoticed: instead of extracting the genome from the cell nucleus, it was possible to isolate a smaller DNA sequence from the mitochondrion—one of the cell's components. This is because the structure of deviant codes (codes deviating from the standard structure) can reduce the size of the entire genome. Mitochondrial DNA is typically recorded using such codes. Such a sequence will contain practically no "passive" (inactive) regions, which would otherwise significantly slow down the DNA identification process.

In 2012, the Nobel Prize in Physiology or Medicine was awarded to experts in cellular biology: the British scientist John Gurdon and the Japanese scientist

Shinya Yamanaka. Their research focused on the reprogramming of genetic information that controls the development of living cells.

Man, using his hands and intelligence, creates new forms of living organisms that often not only match but even surpass those created by nature. The prophetic words of I.V. Michurin come true: "Man can and must do better than nature."

Academician N.P. Dubinin wrote: "To solve the problem of achieving directed genetic changes means to learn how to control life. This is the most fascinating and urgent task of modern natural science. Such a problem will be solved through the joint efforts of genetics, cytology, chemistry, physics, cybernetics, and other sciences. The solution to this problem will place life in the hands of humanity."

Modern mathematics should be considered a combination of deductive and inductive approaches.

A key aspect of justifying the recognition procedures or inductive reasoning methods is the presence of all classes within the sample, as well as averaging the error across multiple training samples [5].

At first glance, it might seem that the methods of biological research (since they are used for empirical knowledge acquisition) can only be of an inductive nature. Indeed, this was precisely the view held by the "all-inductivists" of their time, about whom Engels famously observed that they "got stuck in the contradiction between induction and deduction" [6].

The deductive construction of a biological theory using an axiomatic method assumes that it is possible to identify a number of statements from the entire body of scientific knowledge for which their truth is accepted without the need for proof.

In modern science, concepts such as structure, symmetry, and proportion play a fundamental role, and these originated in mathematics.

Mathematics has successfully tackled the problems that life has posed to it. Mathematics also played a special role in the process of genetic research. A great contribution by G. Mendel was his ability to find an incredibly simple way to express the types of inherited traits observed in pea crossings, as well as their numerical relationships, in mathematical formulas.

However, the development of mathematics has not always been smooth. There have been periods of decline as well as periods of growth in mathematical thought.

Nevertheless, it is precisely the combination of the complex internal structure of mathematics with real-world application problems that provides the driving force behind the advancement of this science.

In their search for methods to model complex systems, researchers turn to various branches of mathematics. The guiding principle for their approach is Gödel's Incompleteness Theorem, which demonstrates that it is impossible to completely formalize all knowledge. This necessitates the use of a combination of both mathematical and heuristic methods in the conduct of systematic studies [7].

Currently, Markov chains are gaining increasing attention in biology and genetics, and together with other types of mathematical models, they occupy a quite prominent position. Numerical calculations conducted using information from the Protein Data Bank (PDB) have confirmed the high efficiency of

Bayesian recognition procedures on Markov chain models in the task of identifying protein secondary structures.

In further research, a significant breakthrough should be the creation of a functional, complete mathematical model of a living cell, as such a model would have predictive capabilities. To develop such a model, the most advanced mathematical methods are required—especially those that can identify and process information in the most optimal manner.

The world's first complete computer model of a living organism has been created by Markus Covert and his colleagues at Stanford University. The model was based on the bacterium Mycoplasma genitalium, which possesses one of the shortest genomes known. Nearly 2000 experimental relationships, reactions, and other characteristics specific to this bacterium at various stages of its life cycle were incorporated into the model. Many phenomena in living organisms are the result of interactions between hundreds or even thousands of genes. In the future, CAD (Computer-Aided Design) modeling of living organisms could become one of the most ambitious projects in science. [8]

The basis for writing this book was the mathematical methods for diagnosing technical systems and genetic information, using Bayesian recognition procedures, as described in the monographs [9, 10] and the articles in the journals [11, 12].

Due to the discrete nature of the genetic code, it is natural to utilize it in the context of modern theories of discrete signals. The content of this material introduces readers to mathematical and logical methods for analysis and recognition. These methods are described in terms of Boolean algebra.

I believe the call for in-depth and productive research in the field of genetics will be heeded. Most likely, over time, the future of genetics will become understandable to the populations of many countries—and, more importantly, accessible, as the cost of genetic testing will decrease significantly. Then genetics will make a breakthrough among the general public, just as information technology has done in recent years. Perhaps in twenty or thirty years, most people will have their own genetic passport.

I would like to express my gratitude to the members of the family: my daughter Elena Vitalievna, my grandson Maxim Andreevich Krupskiy, and my wife Ekaterina Alekseevna Gupal, for their active participation in the preparation of this monograph and the revision of the manuscript.

CHAPTER 1. MODERN MATHEMATICS AS A COMBINATION OF INDUCTIVE AND DEDUCTIVE APPROACHES

1.1. AXIOMATIC METHOD

The axiomatic method of constructing a scientific theory involves accepting certain fundamental statements, known as axioms, "without proof"; all subsequent assertions within the theory are then derived from these axioms through logical reasoning.

The axiomatic method in mathematics was first applied by Euclid in his book "Elements," which presented the foundations of elementary geometry, number theory, algebra, and other branches of ancient mathematics.

Some of the definitions in Euclid's "Elements" are merely descriptions of basic concepts. For example: "A point is something that has no parts," "A line is a length without width," and "A straight line is one that lies evenly with respect to the points on it."

Obviously, such "definitions" are unlikely to be used in mathematical proofs. However, there are also definitions that are considered valid in the modern sense of mathematics. For example: "Parallel lines are straight lines that lie in the same plane and, if extended infinitely in both directions, never intersect."

Following the definitions come the postulates, which assert the possibility of performing elementary constructions.

Following the postulates in Euclid's 'Elements,' the axioms are presented—statements about the properties of the relationships of equality and inequality.

Based on definitions, postulates, and axioms, new geometric statements—theorems—are derived through proof.

Since geometry was supposed to be a description of the real physical space, it is quite natural that Euclid considered concepts such as "point" and "line" to have fairly clear meanings, and the corresponding postulates and axioms to be "self-evident truths."

The construction of Lobachevskian geometry models was of fundamental importance for the development of the axiomatic method, as it led to the realization that the axiomatic theory could be approached purely formally, without assuming any predetermined meaning for the basic concepts. In fact, we can choose the meanings of these concepts in any way whatsoever, as long as the resulting axioms remain true.

In the 19th century, the axiomatic method became widely prevalent in mathematics. D. Peano (1891) proposed an axiomatization for the natural numbers. Axiomatic theories for real numbers were also developed, and eventually, a system of axioms for set theory was established. Formal axiomatic methods have particularly gained widespread use in modern algebra, where the axiomatic system essentially serves to define a particular algebraic structure [13].

1.2. JUSTIFICATIONS FOR THE INDUCTIVE METHOD

It cannot be said that the axiomatic-deductive method was the only approach to studying phenomena and processes in the real world. An alternative to it is the inductive approach, which is a method of reasoning from the particular to the general.

"The mother of all errors and misfortunes in science," wrote Francis Bacon, "is the method of discovery and verification that begins by establishing the most general principles, and only later adapts intermediate axioms to those principles for further testing." According to Bacon, the true path of science is one that "derives axioms from sensations and particulars, gradually ascending to the most general ones. This is the true way, but as yet untried." Bacon outlines and defends the concept of a new, "true" science: its foundation should be based on systematic observation and experimentation, and its method would involve "constructing concepts and axioms through genuine inductive reasoning."

F. Bacon sees the main task as developing rules that would enable a scientist to discover the laws of nature based on observational data. It is in these rules, rather than in the search for the ultimate foundations of nature, that Bacon believes lies "the entire usefulness and efficacy of practice."

The reference to intellectual intuition as a means of discerning truth in geometry, and indeed in other sciences, has been significantly weakened. In fact, Lobachevsky's postulate, which states that an infinite number of lines can be drawn through a point outside a given line and parallel to that line, contradicts all common sense. Moreover, the parallel postulate in elliptic (special Riemannian) geometry claims the exact opposite: no lines can be drawn through a point outside a given line that would be parallel to it. [5]

There were several main alternative approaches to justifying the concept of induction. This diversity of approaches stems from both the ambiguity in the understanding of the nature of induction itself and the varying arguments put forward in defense of the reliability of induction as a particular type of reasoning.

The conclusions of inductive arguments always assert more than what the premises alone allow. Therefore, inductive reasoning is often described as involving an "inductive leap" – the conclusion expands and strengthens the content of the premises.

Since the conclusion of enumerative induction does not logically follow from the content of its premises, a natural question arises: What is the basis for this inductive leap? What principle supports the belief that something will happen simply because it has happened in the past?

J.S. Mill attempted to answer this question within the framework of what is known as a metaphysical or deductive justification of induction. Supporters of this approach sought to justify induction by referring to some universal principle of an ontological nature. By incorporating such a principle as a "larger premise" into the inductive argument, they argued that the inductive conclusion would thereby become a specific form of deductive reasoning.

In fact, information is contained not only in true statements but also in false ones. Therefore, it's not surprising that all attempts by mathematicians to construct a complete deductive system based solely on true statements have been unsuccessful. From a false statement, it is sometimes possible to derive a true one; in other words, a correct conclusion can be reached from relationships between false statements. For example, Aristotle states in the second book of his Premiers Analytiques: "It is possible to obtain truth from false statements, which may themselves be either both false or only one of them false." He demonstrates this using various methods. Let's consider an example where only one of the two premises is false, yet the conclusion is true:

Every horse is an animal.
A human being is not an animal.
Therefore, a human is not a horse.

All the conclusions derived from true sentences are indeed true. However, sometimes the conclusions of false sentences can also be true. To prove that a given sentence is false, it is sufficient to show that, through valid logical reasoning, a false sentence or one incompatible with one of the preceding ones can be derived from it.

To determine whether a sentence is true, one must proceed in the manner one would if wishing to prove that it is true; if this is successful, the question is resolved. However, if no sentences that are recognized as true can be derived from the given sentence, then no positive conclusion should be drawn from it—such a result merely proves the powerlessness of the attempt. In such a case, one should attempt to determine whether the sentence is false. To do this, one proceeds in the manner one would if wishing to prove that it is false, that is, trying to derive some false sentence from it as a necessary consequence. If this is successful, the question is resolved; otherwise, we must remain in doubt.

Therefore, in order to determine whether a statement is true or false, one must refer back to the questions that have already been discussed and whose purpose is to prove the truthfulness or falsity of that statement [14].

1.3. THE CONTRIBUTION OF K. GÖDEL TO INDUCTIVE MATHEMATICS AND LOGIC

Gödel presented his famous "Incompleteness Theorem" in October 1930 at a meeting of one of the sections of the Vienna Academy of Sciences. The article containing a detailed exposition of the theorem was published in 1931. In 1940, Gödel moved to the United States, where he served as a professor at the Institute for Advanced Study in Princeton for many years. For his work in cosmology (the "Gödelian universe" with closed time lines), he was awarded the Einstein Prize in 1951. Gödel died in the United States in 1978. [15].

"The Incompleteness Theorem" is a mathematical assertion made regarding logic, the science that studies methods of proof and refutation. However, the significance of Gödel's theorem is not limited to mathematics alone; experts from various fields of knowledge utilize logical frameworks when constructing any scientific model. Philosophers, physicists, psychologists, biologists, and specialists in other disciplines also refer to this theorem.

In the paper that contains the proof of the "Incompleteness Theorem," Gödel studied a system of normal arithmetic of the type presented in "Principia Mathematica" (PM).

Gödel then goes on to outline the main idea of his proof, which cannot be assessed in any other way but as brilliant. He decides to use a concept that previously played a purely negative role in the foundations of mathematics in order to achieve a positive logical-mathematical result. Specifically, he constructs his proofs based on the famous "Liar" paradox, which has been known since ancient times. This paradox is described in various forms in different ancient sources and is attributed to Eubulides (4th century BCE). The simplest and most expressive form of this paradox goes as follows: Someone says, "I am lying." What was stated—truth or falsehood?

Implementing the reasoning conceived by Gödel required thorough preparation.

According to Gödel's theorem, statements in a particular logic will always fall into one of three types:

1) Statements derived from axioms are true.
2) Statements that contradict the axioms are false.
3) Statements that fall into neither of the above categories.

Gödel's proof is nothing more than the realization of a practical idea: by finding new axioms, we ensure that there is no ambiguity left whatsoever. In other words, statements can be unambiguously classified as either true or false.

Gödel represents the formula within the coding system he developed; this formula asserts its own unprovability. If the logical-arithmetical system in question is consistent (i.e., all formulas provable within it are true), then the statement "I am unprovable" becomes unsolvable. This constitutes Gödel's main result: in a formal arithmetic of the "PM type," such a statement is neither provable nor refutable (in other words, its negation cannot be proven).

The incompleteness of the system indicates that the system lacks the necessary tools to "model" the concept of truth within that particular domain.

According to the "Incompleteness Theorem," if a formal arithmetic system (such as the type PM) is consistent (i.e., free from contradictions), then it is incomplete. What if the system is contradictory? In that case, the very notion of a "theorem" loses its meaning within such a system, because any formula can be proven in a contradictory system. A formula that claims "I am unprovable" would then be provable; likewise, its negation would also be provable. Mathematicians and logicians believe that the theory of natural numbers is consistent (i.e., free from contradictions).

Gödel's results apply not only to formal arithmetic but to any formal system that contains arithmetic, that is, to calculations "that start with arithmetic."

Gödel's discoveries initiated rigorous investigations into the possibilities of the formally deductive method of knowledge. These findings were almost entirely unexpected for the logic community prior to Gödel's work. However, a common and quite compelling motif underlying them is the conclusion that there are inherent limitations to regular deductive and computational procedures; in other words, it is impossible to represent the process of expanding knowledge (beginning with mathematics) as a complete, formal system.

The main conclusion from the "incompleteness theorem" can be stated as follows: Any comprehensive, but rigid (unflexible) system of thought is inevitably imperfect; such a system will contain problems that cannot be solved using that system itself.

It is precisely in the rigorous demonstration of the fundamental imperfection of any rigid, frozen system of thinking that the true dialectical significance of Gödel's achievements lies. Imperfect axioms must be improved upon, and this is indeed the case in mathematical practice ([16, 17]).

In March 1956, K. Gödel wrote a letter to his friend John von Neumann about a mathematical problem. He wondered how long it would take for a hypothetical machine to provide answers to the problem and went on to say, "If such a machine were actually created, the consequences would be of tremendous significance..." However, von Neumann was seriously ill and died less than a year later, never responding to Gödel's letter. Nevertheless, the problem outlived both of them. Gödel's question, now known as the $P_{vs}NP$ problem, has become the foundational concept behind modern computers. It gave rise to a new theory called computational complexity theory, which combines mathematics, science, and engineering in an effort to determine with complete certainty what computers can and cannot do in real-world situations.

The $P_{vs}NP$ problem is the question of whether complex problems whose solutions can be quickly verified can also be solved quickly [18].

1.4. THEORY OF COMPLEXITY FOR ENUMERATION AND COMBINATORIAL PROBLEMS

At the end of the last century, a theory of the complexity of enumeration and combinatorial problems was developed. All the recognizable problems for which a polynomial-time algorithm exists form a class called P. This name originates from the concept of "polynomial time," which is a mathematically precise way of expressing the fact that, no matter how long a list of items (for example, product prices) may be, the time required to sort it (arranging the prices in ascending order) will never increase uncontrollably. A problem is considered difficult if there is no polynomial-time algorithm to solve it. In practice, however, there are some problems whose solutions can be easily verified; experts refer to such problems as NP-problems. Currently, the most complex problems in the NP-class (so-called NP-complete problems) can only be solved in exponential time, which is considered unacceptable from a practical standpoint. For instance, the simplex method for solving linear programming problems has an exponential time complexity, but numerous studies have shown that it works well in practice. It turns out that many well-known problems belong to the category of NP-complete problems, and the list of such problems exceeds one hundred. To this day, no polynomial-time algorithm has been developed for any NP-complete problem, regardless of the size of the input data. In 1971, the Canadian mathematician S.A. Cook put forward a hypothesis, and in 1972 the American mathematician R.M. Karp proved a theorem on the reducibility (equivalence) of all NP-complete (i.e., universal) enumeration (or combinatorial) problems to one another. The hypothesis states that either all NP-complete problems have an algorithm whose running time is polynomial in the length of the input, or none of them do. Although the NP-completeness problem remains unsolved, many experts in complexity theory believe that it will be solved eventually. If a fast solution to an NP-complete problem is found, then all NP-problems can be solved quickly. This would mean that class P would be equivalent to class NP.

Constructing a polynomial-time algorithm for any NP-complete problem on a deterministic Turing machine implies that such algorithms can be automatically constructed for all other NP-complete problems as well.

In 1979, in the book [19], American mathematicians M.R. Garey and D.S. Johnson developed the theory of NP-completeness and gave formulations of three hundred NP-complete problems.

However, it is also possible that the hypothesis that $P \neq NP$ (which is accepted by many mathematicians) cannot be proven or refuted.

In practical work, a person was interested in obtaining precise solutions to problems, as the effectiveness of the decisions made depended on the accuracy of those solutions. In the 20th century, solving complex problems required powerful computing resources and significant amounts of computer memory.

Achieving an exact solution and providing a rigorous proof became an insurmountable barrier for mathematics. To make progress, even in just part of solving NP-complete problems, a decisive step is needed: to abandon the requirement for an exact solution and introduce the concept of "approximate solutions" into mathematics.

In fact, the root cause of the crisis in mathematics was the widespread emergence of enumeration (search) problems. The inadequacy of mathematical logic and the foundational principles of mathematics to address these combinatorial issues necessitated the search for non-traditional approaches. This in turn led to the need for developing logical tools, information representation methods, and algorithms for processing data that were different from those inherent in classical mathematical logic and its foundations.

Geneticists suggest that the total number of genetic codes, or their diversity, is estimated to be around 10^{1000}. "The selection of a particular genetic code throughout the entire history of the universe—whose duration, according to scientists, is 13 to 15 billion years—must have been accomplished in some incredibly 'unusual' way, as a complete enumeration of all possible codes seems impossible. Consequently, there must exist other mechanisms or approaches for the generation of genetic codes." [20]

In 2000, in his dissertation, V.Z. Strygin demonstrates the fundamental possibility of solving NP-complete problems, in particular by utilizing a new mathematical model known as matrix logic. This information enables many researchers in various fields of science and technology to search for (and develop) polynomial-time algorithms to solve their combinatorial problems. rather than being confined to indications of the fundamental irreducibility of enumeration in certain enumeration problems, and to the widespread belief that all NP-complete problems are indeed intractable—meaning they do not have polynomial-time algorithms. [21]

The analysis of samples in classification tasks, particularly those related to finding a logical function that divides the input set into classes, requires performing a massive amount of iterations using methods for minimizing Boolean functions. A way to deal with poorly formulated or ill-defined problems is through directed (guided) enumeration. Currently, so-called evolutionary algorithms are widely used for these purposes. Evolutionary algorithms can be classified under the well-known trial-and-error method, associated with solving enumeration problems that arise when searching for correct solutions or extreme values, in cases where there is no knowledge or theory about the behavior of the system or process under study (Reference [22]).

Extreme Combinatorial Problems (ECPs) attract the interest of physicists, biologists, mathematicians, economists, sociologists, and engineers due to their relevance to modeling physical, technical, and technological processes.

A significant characteristic of biological research is its predominantly empirical nature. Since observations and experimental research are decisive in biology, the purely logical development of theoretical predictions plays a subordinate, auxiliary—though ever-expanding and strengthening—role here [23].

At first glance, it might seem that the methods of biological research (since they are used for empirical knowledge acquisition) could be purely inductive. Indeed, this was precisely the view held by the "all-inductivists" of their time, about whom Engels famously observed that they "got stuck in the contradiction between induction and deduction" [6].

The deductive construction of a biological theory using an axiomatic method assumes that it is possible to identify a number of statements from the entire body of scientific knowledge for which their truth is accepted without the need for proof.

Furthermore, all other knowledge is deduced logically from these initial axioms or logical principles. Thus, the scope of logical conclusions is confined to the realm of true statements. In fact, this limitation is quite significant; it essentially results in the low efficiency of the deductive reasoning process. From the perspective of information theory, this means that useful information contained within false statements is discarded, thereby restricting the capabilities of deductive procedures.

Induction is a logically reverse form of deduction. The process of selecting premises for induction itself is already deductive in nature; at the same time, the main premise of deduction is, more often than not, arrived at inductively. Induction begins with the conclusions that deduction ends with, and vice versa. The necessary connection between induction and deduction, as well as their dialectical unity, is also reflected in the fact that both are cognitive processes that regulate each other [25]. Therefore, modern mathematics should be considered a combination of deductive and inductive approaches. The domain of deductive reasoning is the derivation of precise solutions to problems of small scale; the domain of inductive reasoning is the solution of large-scale problems based on approximate calculations.

CHAPTER 2. THE PHENOMENON OF LIFE ON EARTH

As humans gain knowledge, they delve deeper into the mysteries of nature and unravel its age-old secrets.

The ancient Greek philosopher Protagoras argued for the subjective nature of knowledge and proposed the statement: "Man is the measure of all things…" [24]. From this, it follows that man also serves as the measure of truth.

The human being is a unique creation of the universe, possessing a number of extraordinary qualities, most of which have yet to be fully understood or utilized by us.

According to Pierre Teilhard de Chardin, the French biologist and thinker, the human being is the key to all science and nature; he represents the central point of perspective that summarizes everything we know. To understand the human being is to understand the past and the future of the world.

It's time for humans to realize who they are and what their role is in the universe.

Life is one of the forms of existence of matter that arises systematically under certain conditions during its development.

Science has established that the basis of life lies in organic substances: deoxyribonucleic acid (DNA), ribonucleic acid (RNA), and proteins. DNA is the hereditary material of the cell. Today, DNA and RNA molecules are considered to be the physical carriers of biological information (memory). Scientists believe that a versatile molecule like RNA played a crucial role in the emergence of life, while DNA emerged later in the course of evolution. RNA is regarded as a sort of "ancestor" of DNA, having long since transferred all its regulatory functions to its younger and more successful "descendant."

The view that the RNA molecule is the primary molecular entity that constitutes life is supported by the following reasoning:

1) RNA carries genetic information.
2) Viral RNAs are capable of recombination involving both viral and cellular RNAs.
3) RNA possesses catalytic properties.

If we consider that life is impossible without DNA and RNA, then it must be acknowledged that the phenomenon of life on Earth only emerged when our planet produced its first (and functioning) molecule of this type. Attempts to mathematically estimate the likelihood of such a complex biomolecule spontaneously forming in a "broth" containing all the necessary chemical components have yielded a vanishingly small value. As a result, even the most staunch proponents of strict materialistic views are now inclined to consider the possibility of life originating from outer space—perhaps through the delivery of DNA molecules via meteoritic material (e.g., [26]).

Let's consider some hypotheses about the origin of life on Earth. Our planet has existed for approximately 4.5 billion years. According to the majority of scientists, life emerged spontaneously as a result of random chemical reactions that, over millions of years, transformed simple substances contained in Earth's waters into living organisms. It is likely that certain reactions produced compounds with the ability to replicate themselves. However, there are also other hypotheses regarding the origin of life on Earth; these hypotheses are diverse and have not yet been definitively proven. Different hypotheses have been proposed at various times and in different cultures. The main ones include:

- Creationism (from the Latin word "creatio" – meaning creation) is a religious doctrine that explains the origin of the world as being created by God from nothing. It is characteristic of theistic religions such as Judaism, Christianity, and Islam.
- Spontaneous origin: It is believed that life has arisen multiple times from non-living matter.
- The hypothesis of a static state suggests that life has always existed.
- Cosmic origin: Life originated from other planets and was brought to Earth via panspermia.
- Biochemical theory: Life originated under Earth's conditions through processes that obey physical and chemical laws, that is, as a result of biochemical evolution.

Today, biologists only accept the last option as a scientific hypothesis. Panspermia also does not contradict the fundamental principles of science; however, it is not considered an independent hypothesis in its own right, as it merely transfers the process of the biochemical origin of life to other, extraterrestrial environments.

Let's take a closer look at the listed hypotheses.

Creationism: According to this religious concept, which has ancient roots, everything that exists in the universe, including life, was created by a single Power—the Creator—as a result of one or more acts of supernatural creation in the past.

The concept of the origin of life cannot be either definitively proven or refuted, and it coexists with various scientific theories regarding the origins of life.

The hypothesis of spontaneous generation: For thousands of years, people believed in the idea that life could arise spontaneously. However, by the late 1870s, nearly all scientists agreed that living organisms could only come from other living organisms. This led to the re-emergence of the original question: Where did the first organisms come from?

The hypothesis of a static state: Based on this hypothesis, the Earth never came into existence but has always existed. It has always been capable of sustaining life, and if it has changed at all, then only very slightly. This hypothesis is not considered to be scientific.

The Panspermia Hypothesis: The hypothesis that life on Earth originated as a result of the transfer of certain life forms from other planets is called Panspermia (from the Greek words "pan" meaning "all, every," and "sperma" meaning "seed"). According to this hypothesis, life has existed throughout the universe forever, and these life-form seeds travel across the cosmos, being transported from one planet to another. It is possible for them to reach our planet as well. This theoretical possibility is supported by the discovery of traces of organic compounds in meteorite and comet materials.

In the article "The Galaxy as a Biosphere" (reference [8]), Alexey Levin notes: "The origins of the panspermia idea can be traced back to early antiquity, but it only flourished in the 19th century…" This theory was later supported by Francis Crick, one of the discoverers of the double-helical structure of DNA and a Nobel Prize-winning British biophysicist, as well as Fred Hoyle, who coined the term "Big Bang". Currently, this idea is being developed by Nalin Chandra Wickramasinghe, director of the Buckingham Center for Astrobiology in the UK, and his colleagues. Wickramasinghe's model, which they call "Hydrogravitational Dynamic Cosmology," predicts that life will inevitably emerge not on individual planets, but throughout the galaxy as a whole—thanks to complex molecules that are carried by comets and other "roaming" celestial bodies.

Francis Crick and Leslie Orgel, in considering the research on the origin of the genetic code published at the end of the last century, believe that the complex molecular mechanisms required for translating RNA into proteins could have been brought to Earth by living organisms (bacteria) from some other part of the universe.

Fred Hoyle and Chandra Wickramasinghe provide compelling evidence in their works that many objects in the Solar System—especially comets, as well as planets with organic atmospheres (containing ammonia and methane)—are rich in cellular forms of life. The early stages of chemical evolution may not have taken place on Earth. [27]

According to this hypothesis, genomic information extends beyond the boundaries of chromosomes and even beyond the Earth itself; it exists on a universal scale, fertilizing all new worlds whose physical and chemical conditions are suitable for the emergence of life in some form or another.

Biochemical Theory: The first scientific theory regarding the origin of living organisms on Earth was proposed by the Soviet biochemist A.I. Oparin (1894–1980). The theory of biochemical evolution has the largest number of supporters among modern scientists.

Generally, chemists studying natural compounds follow this logic in their work: they first discover a new substance in nature, then determine its functions and structure, and finally attempt to synthesize it in the laboratory in order to compare the properties of the natural compound with those of its synthetic counterpart. Only in this way can it be proven that a substance with a particular chemical structure indeed possesses certain properties. However, this approach did not work effectively for genetic manipulations for a long time. The structure of DNA was already known, but the reverse problem could not be solved [28]. Living organisms seemed so complex that many biologists initially believed life could not be reduced to simple chemical reactions; nevertheless, in 1828, Wöhler successfully synthesized an organic molecule chemically—urea, which was obtained from inorganic substances. It was not until 1953 that Stanley Miller synthesized the components of proteins, namely amino acids [29].

In the 1920s, A.I. Oparin proposed a theory about the origin of life on Earth. Most of the surface of the primordial Earth was covered by a layer of hot water, the boiling of which was maintained by magma located beneath the thin oceanic crust. Such a mixture of gases and hot water could have led to the formation of what is known as the "primordial soup," rich in the chemical elements necessary for the synthesis of life. The reaction could have occurred as a result of volcanic activity, intense ultraviolet radiation passing through the thin atmosphere, or an electric discharge from a lightning bolt.

This theory was experimentally tested in 1953 by the American scientist Stanley Miller. Miller created a model of the primordial world consisting of two flasks and glass tubes. One of the flasks contained a solution whose composition theoretically matched that of seawater. The space above the liquid was filled with a mixture of gases that, again, theoretically corresponded to the supposed atmosphere of that time. This flask was connected to the other flask via a tube that contained two electrodes used to generate a spark—a miniature model of a lightning bolt. Another tube extended from this "spark chamber" and led back to the first flask, passing through a capacitor and a P-shaped collector.

When Miller heated the mixture in the lower flask, it boiled and turned into a gas. The gas then entered the spark chamber, where it condensed and flowed back into the lower flask. This process was continuously repeated for a week, after which the liquid was extracted for analysis. The resulting mixture contained three amino acids—compounds from which proteins are formed. Many researchers adopted this approach, conducted similar experiments, and discovered even more amino acids as well as simple nucleotides, which are the building blocks of DNA. The results of these experiments suggested that it is possible for all proteins (and perhaps other substances as well) to have been synthesized over billions of years. Many scientists are still searching for the mechanism by which amino acids can be spontaneously assembled into proteins without the direct involvement of DNA. [30]

The experiment to create artificial life also involves the following steps: scientists synthesize the genome of one bacterium and insert it into a cell of another bacterial species. The resulting organism, which has the outer membrane of the recipient bacterium (Mycoplasma capricolum), turned out to be identical to the donor bacterium (Mycoplasma mycoides). This experiment provided the first conclusive evidence that DNA indeed contains all the information necessary for the functioning of a living cell.

V.M. Nosov states in his forecast [31]: "The process of the emergence of life in a star system and its development until the appearance of intelligent beings takes approximately 4.5 billion years from the moment the star system is formed. This period of time was sufficient for life to arise on Earth and to develop until the formation of intelligent communities." He further argues that life emerges in the planetary systems of all sun-like stars. In most of these types of stars that we observe, civilizations—intelligent beings—have emerged and are currently developing successfully. Our universe, as determined by scientists, has existed for 13 to 15 billion years. Since it has been evolving throughout this time, life has continuously arisen in various places within our starry world, and new intelligent communities continue to form.

V. I. Vernadsky, recognized as a classic figure in the field of natural sciences, uses the term "geomerids" as a synonym for what is often referred to as the "living covering of the planet." He views these geomerids within the context of biogeochemistry. The habitat of geomerids is the Earth's landscape layer. In the broader concept of life, this encompasses organisms that exist not only on Earth but also on other planets. It was the geomerids that "enlivened" the originally lifeless landscapes. Thus, the living nature of our planet came into being. [32]

There are examples in the world of seemingly inexplicable interventions in the evolution of humanity from external sources. One such phenomenon is the civilization of Sumer, which existed more than three thousand years before our era. The Sumerians had precise calendars and understood the structure of the universe; they even accurately depicted the layout of the solar system. Sumerian doctors were considered to be highly skilled healers. However, it remains unknown where this people came from—something quite unusual for that time period. Moreover, no genetic connections have been established between their language and other languages.

Modern science is still far from providing a comprehensive explanation of how inorganic substances have developed to such a high level of organization, characteristic of biological processes. Reconstructing the specific mechanisms behind this gradual increase in complexity will be the task of future scientific research.

On October 4, 1957, with the launch of the world's first artificial satellite in the Soviet Union, a new phase in the development of human civilization began. This marked a real step towards exploring the universe in search of other inhabited regions by intelligent life forms.

The universe is permeated by a kind of internal "shaking," similar to the hissing noise produced by electrical static in a short-wave radio receiver. "If we really pursue this line of thought further," says Craig Hogan, director of the Center for Elementary Particle Astrophysics, "we will conclude that the digital noise inherent in space can be measured." If Hogan's discovery is confirmed, it would mean that the universe is indeed "digital" in nature. It turns out that the phenomena occurring at this level are fundamentally related to information theory—the zeros and ones that constitute the basis of the universe.

In the second half of the last century, a systematic search for extraterrestrial life was initiated under the SETI program, which involved monitoring the radio sky in search of signals from alien civilizations.

Recent research by scientists indicates that, in order to survive, cells exchanged useful components with each other and thus formed a global mega-organism. Three billion years ago, this mega-organism split into three distinct domains of life: single-celled bacteria, archaea, and the more complex eukaryotes, which gave rise to all animals and plants.

So, what exactly is LIFE?

Life is one of the most complex phenomena in nature—a special, highest form of material movement that emerged at a certain stage in its historical development and is characteristic of complex protein structures. The first comprehensive scientific definition of life was given by Engels: "Life is a mode of existence of protein substances, and this mode of existence consists, in essence, in the constant renewal of their chemical components through nutrition and excretion." [33] Is it possible to decipher all proteins, understand their structure, functions, and thereby gain knowledge of ourselves? Will this philosophical question remain eternal?

The characteristic that defines living things is the incredible diversity of life on Earth: in terms of shape, appearance, size, organization, diet, living conditions, and many other aspects. However, as our knowledge has deepened, we've realized that all these are merely "external manifestations" of more fundamental properties. At the molecular level, the diversity of living organisms reveals a completely different nature. Unlike machines and mechanisms, living matter and humans cannot be disassembled into their constituent parts without causing catastrophic consequences; such disassembly would prevent us from accurately identifying any potential defects. Machines and mechanisms, on the other hand, can be broken down into individual components. The difference between machines and living organisms is also highlighted by the fact that cellular organelles (such as ribosomes and chromosomes), as well as the organisms themselves (like hydras and slime molds), are capable of self-assembling. This is impossible for machines. Living organisms are distinguished from inanimate objects by their ability to move, digest food, and contract their muscles. These complex functions are carried out by protein-based "machines," and the final structure of protein molecules is determined by the balanced distribution of electrical charges along their amino acid chains.

The ability to create complex structures from simpler ones is characteristic only of living organisms. No matter how "living" a robot may appear, it is merely a machine that operates using energy. However, robots cannot obtain that energy on their own (unlike living beings) and are therefore entirely dependent on humans. Additionally, robots do not have the ability to grow or reproduce.

Materialist scientists have proven that the essence of life, its material basis, consists of living protein structures. These structures are characterized by a constant exchange of matter and energy with the surrounding environment. The processes involved in this exchange are assimilation and dissimilation. These processes occur continuously and simultaneously, in close interconnection. The compounds that are destroyed are replaced by newly absorbed substances, thereby enabling the gradual renewal of the organism.

The ancient Greek scientist Heraclitus said that our bodies flow just like rivers, and that matter is constantly renewed within them, much like water in a stream. Life itself can only exist under conditions where proteins are able to exist as chemical entities, and where there is an exchange of substances between proteins and the surrounding environment [34].

The evolution of living organisms began with the emergence of cells and single-celled organisms.

Cell theory was first clearly formulated in 1839 by German researchers Schleiden and Schwann. This theory states that all plants and animals are composed of small, elementary subunits called cells.

A schematic illustration of a plant cell is shown in Figure 2.1.

Cells are only formed from other cells during the process of cell division. During this process, the cell nucleus also divides, and both daughter cells receive a nucleus each. The nucleus contains genetic material. The cell nucleus contains several molecules of DNA; each molecule of DNA forms a chromosome. In a somatic cell, there is a pair of identical chromosomes—one from the mother and one from the father.

The synthesis of protein molecules (the primary "working mechanisms of life") occurs within specialized molecular machines known as ribosomes. The transfer of information from DNA to ribosomes is facilitated by a special type of macromolecule: messenger RNA (mRNA). The isolated nucleic acids and proteins are capable of recognizing each other; upon interacting, they can assemble into a complete and fully functional ribosome. Proteins vary in structure, shape, and size, but all of them contain functionally unified "working" elements [35].

The main products of enzymes are the same for all living organisms: sugars, amino acids, fatty acids, etc. DNA ensures the synthesis of identical proteins, regardless of how it is transferred. Similarly (due to its universality), structural proteins also behave in a similar manner; they possess regions that can interact with each other.

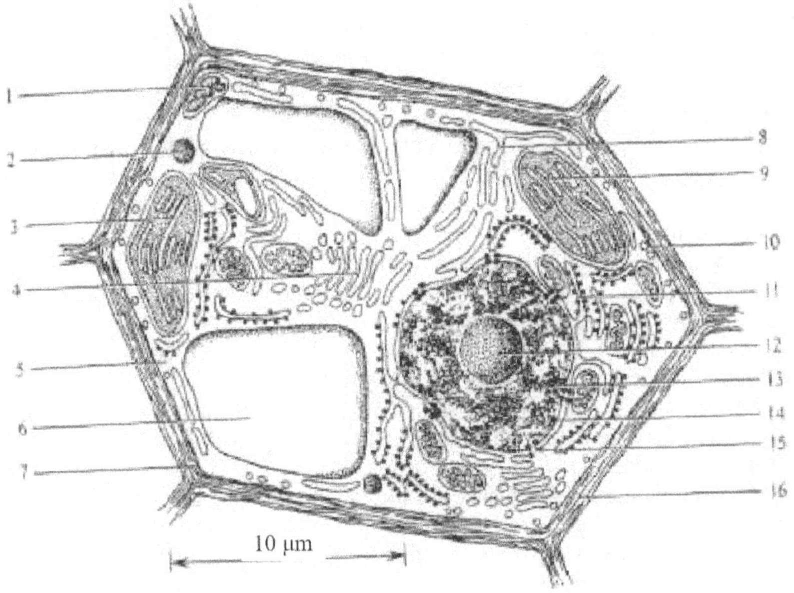

Figure 2.1. Schematic illustration of a plant cell

1 – Mitochondrion; 2 – Living inclusion; 3 – Chloroplast; 4 – Golgi apparatus; 5 – cytoplasm; 6 – vacuole; 7 – Plasma membrane; 8 – Endoplasmic reticulum (smooth); 9 – Chloroplast; 10 – Microtubules; 11 – Endoplasmic reticulum (rough); 12 – Nucleolus; 13 – Condensed chromatin; 14 – Nuclear membrane; 15 – Nuclear membrane pore; 16 – Cell wall.

A gene is a segment of a chromosome that contains the instructions for the synthesis of a particular protein. In this way, DNA enables the cell to produce the proteins it needs in order to function properly. During this process, the double helix structure of DNA unwinds temporarily, and its genetic information is "transcribed" into another molecule. This copy is then transported from the nucleus to the cytoplasm, where it serves as a template for the synthesis of the corresponding protein (an enzyme). The DNA molecule is extremely long and can often become damaged due to external factors or errors in the cellular mechanisms that regulate genetic material. To maintain cellular viability, all cells must possess repair systems that minimize DNA damage. In most cases, such damage is automatically repaired; however, sometimes it results in permanent changes to the genetic material—mutations. DNA is located in the nucleus, but the instructions it encodes are copied and distributed throughout the cell.

Mitochondria are the cell's energy-producing "factories." All living cells have the ability to convert energy from the surrounding environment into a common intermediate substance: adenosine triphosphate (ATP).

Each nucleus contains a certain number of chromosomes. Before cell division, each chromosome is replicated, resulting in two new chromosomes that are identical to the original one. This process (first observed by Flemming in

1879) leads to an increase in the number of chromosomes within the nucleus. During cell division, the two identical chromosomes in each pair are distributed to the two daughter nuclei. As a result of these events (now collectively referred to as mitosis), each daughter cell receives a set of chromosomes that is exactly the same as the set in the parent cell [36].

Cells must have a membrane that encloses the cell's interior space, and they usually contain a nucleus, which is also surrounded by a membrane.

The components of cell membranes include phospholipids, glycolipids, glycoproteins, and proteins. The structural basis of a biological membrane is typically a double layer of phospholipids, which is embedded with peripheral and integral proteins. Transport systems are integrated into the cell membrane; these systems consist of ion pumps as well as molecular carriers. Membranes enable the maintenance of a specific composition of the environment within the various cellular compartments (such as the cytoplasm).

For ions and hydrophilic substances, the membrane acts as a molecular sieve. The larger the size of the particle, the lower the permeability of the membrane for that substance.

The schematic diagram of a biological membrane is shown in Figure 2.2. Inside the membrane, there is genetic material in the form of DNA, as well as a protein synthesis apparatus containing a central structure known as the ribosome. Additionally, there are molecules that transfer information from the DNA to the central part of the protein synthesis apparatus. Proteins are transported to the membranes and pass through them. The ability of protein structures to fold and unfold is a characteristic feature of all cells. The configuration of protein molecules is determined by their primary structure; however, in most cases, the native conformation does not arise spontaneously. Chaperones are required for this process to occur.

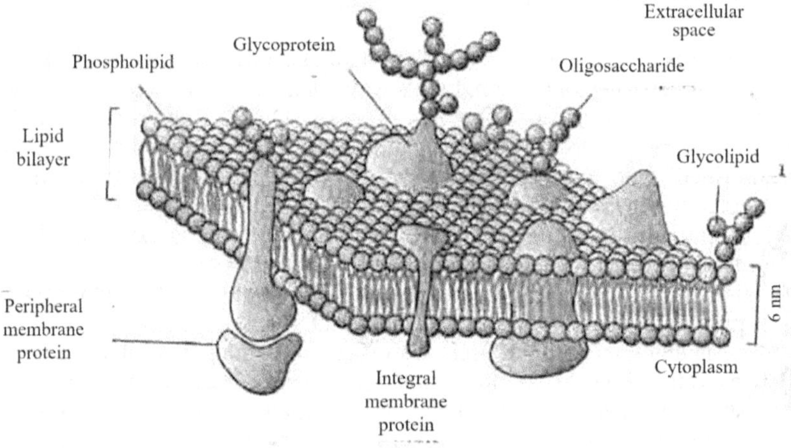

Figure 2.2. Schematic diagram of a biological membrane

Molecular chaperones bind to proteins and facilitate the formation of their native structure. The driving force behind the establishment of a protein's native structure is hydrophobic interactions. Hydrophobic domains have the tendency to associate with each other rather than remain in an aqueous environment. However, these interactions can sometimes occur incorrectly, resulting in a defective native protein structure or the aggregation of proteins with other molecules. These structural rearrangements may continue until the protein achieves its correct native configuration. Chaperones are highly active and constitute the core of the protein structure regulation system.

An important factor is the localization of cellular structures. The localization of cellular structures in specific areas reflects a portion of the cell's genetic information. Positional effects play a crucial role during the early stages of an organism's development.

Cells can interact with each other by exchanging substances for purposes of plastic and energy exchange. However, there are also other ways for cells to "communicate" in order to transmit information. There are two main mechanisms for transmitting information from one cell to another. The first mechanism involves the use of electrical signals, which can be considered a type of Morse code. In this case, short electrical impulses are sent from one cell to another, altering the function of the receiving cell. The second mechanism relies on the use of special signaling substances produced by one cell; these substances travel to other cells and affect their activity. However, signaling substances only have an effect on cells that contain specific receptors capable of recognizing them. Information transmission via signaling substances occurs in two ways: First, a regulatory substance produced by one cell (called an "informon") affects neighboring cells. Second, informons are released from gland cells and reach the target cells through the bloodstream; this process is known as humoral communication (from the Latin word "Humor" meaning liquid), and these signaling substances are called hormones [37].

DNA, proteins, and polysaccharides, as the primary biological polymers in cells, are typical polyelectrolytes whose properties are determined by the balance between electrostatic and hydrophobic interactions.

The key biomolecules that enable life processes are not only capable of self-organization (forming the double helix structure of DNA) but also of recognizing other molecules and particles. A classic example of molecular recognition is the immune response. When foreign substances (antigens) enter an organism, the body begins to produce specialized proteins (antibodies) that are designed to specifically bind to and neutralize those antigens [38].

There are quite definite patterns in the selection of atoms that make up cells in living organisms. In 1945, Frey-Wisling discovered the precise positions of the chemical elements necessary for plants within the Mendeleev periodic table. It was observed that the elements required for the construction and function of cells are distributed in the Mendeleev periodic table in a very specific manner; these are the light elements that form a series ranging from carbon to argon. The Mendeleev periodic table is based on the periodic repetition of the configuration of atoms' outer electron shells. The position of an element in the table is related to its chemical and many physical properties. Almost all of modern chemistry is founded on the principles of the periodic law.

However, any complex organic substance is not yet considered life. For inorganic matter to become living, it must go through at least four additional stages of evolution: the self-assembly of molecules, the emergence of membranes and pre-cellular organization, the development of a mechanism for heredity, and finally the formation of a cell.

Everything that is alive functions on the principle of "self-…": self-assembly, self-organization, self-preservation, self-sustenance, self-sufficiency, self-development, and so on. [35].

"Self-assembly" is a term coined by biochemists to describe the spontaneous aggregation of multimeric biological structures. The self-assembly of a protein is encoded in the spatial configuration of its amino acid sequence. Not only ribosomal enzymes and viruses, but also cell membranes are capable of undergoing self-assembly driven by internal molecular information.

The polypeptide chain possesses an inherent property: through a process of self-assembly, it transforms the one-dimensional information represented by the sequence of amino acid residues into a three-dimensional molecular structure. This is an inevitable and automatic consequence of the tendency of water molecules in their surroundings to assume a state of maximum entropy, while the polypeptide chain itself adopts a state with minimum free energy. As such, self-assembly is both necessary and spontaneous, thanks to the chemical properties of water (the main component of all cells) and macromolecules. The characteristics of this self-assembly process are as follows: it is spontaneous, inherent in the molecules, inevitable, hierarchical in structure, and occurs automatically.

At the time when efforts were being made to decode the genomes of living organisms, including humans, an American named Craig Venter participated in the development of a new sequencing method known as "shotgun sequencing."

Another idea of Venter was to create a viable organism with a minimal set of genes. Solving this problem is essential for understanding the origin of life on Earth, which includes studying the pathways of genetic evolution and the mechanisms by which genomes arise. Such a genetic unit could quite simply be called a "life element" – the smallest possible cell. The goal is to create a cell that contains only the genes necessary to sustain life in its most basic form, that is, a "minimal genome." The estimated minimum set of genes (according to recent calculations by Venter's team) ranges from 310 to 388 genes and should include the following essential genetic systems for microorganisms: genes involved in translation, replication, repair, and transcription; genes that regulate anaerobic metabolism; genes for lipid biosynthesis; genes related to protein transport mechanisms; a set of genes responsible for metabolite transport; as well as a complete set of genes for nucleotide utilization and genes for their biosynthesis. By analogy with chemistry, the simplest unit in that field is the hydrogen atom. Although a "minimal cell" has not yet been created, an organism with a synthetic genome already exists and reproduces in the laboratory at Craig Venter's institute. It is a common bacterium that differs from others only in that its DNA was synthesized in a test tube [28].

A bacterium is the most primitive, self-sustaining cell. It consists of a gelatinous membrane filled with liquid, which converts simple chemical substances (hydrogen, oxygen, carbon, and nitrogen) into complex organic compounds such as proteins and carbohydrates. These processes within the cell are known to be governed by an organic substance called deoxyribonucleic acid (DNA). DNA also possesses another important property: it is capable of replicating itself.

Each DNA molecule resembles a spiral staircase, where chains of atoms form the "ramps" (the sides of the stairs), with "steps" located at equal intervals. The entire molecule can split if necessary, with the "steps" separating in the middle. After splitting, the shortened "steps" reassemble by attracting other substances; these substances then combine to form the missing halves of the staircase. As a result, one spiral molecule becomes two. This simple mechanism is fundamental to life. For instance, a single-celled organism grows and reproduces by splitting in the middle and replicating its internal chemical processes.

Today, we have a very detailed understanding of how a cell is structured, but this understanding is static. The 21st century poses the challenge of determining how these components interact with each other in space and time to form complex, dynamic biological systems.

CHAPTER 3. RECOGNITION OF COMPLEX BIOLOGICAL SYSTEMS

3.1. THE ROLE OF INFORMATION IN THE RECOGNITION PROCESS

The twentieth century and the first decade of this century are referred to in various ways as periods in human history, based on the specific level of development achieved during that time: the Age of Electricity, the Age of Atomic Energy, the Age of Cybernetics, the Age of Genetics, and so on. However, the most appropriate designation for them would be the "Information Age."

This is due to the fact that, at the current rate of development of the main information flows (the internet, mobile and intelligent systems, electronic computing devices, and control systems), these have surpassed all other characteristics from the past period. The amount of information available to science doubles every 7–10 years.

In the early 20th century, as the role of information in social activities grew, so did interest in the nature of this concept. The term "information" is used in two meanings: qualitative (concrete) and quantitative (abstract). Modern information technology increasingly relies on biological systems to search for resources and to develop more advanced computer devices. Life itself can be seen as a form of "self-sustaining information," realized through the materialization of such information. In genetics and biology as a whole, it is impossible to describe phenomena in living organisms without considering the role of information.

A "living program" is information that is organized in a particular way (in the form of a record on a medium that is compatible with the systems for its storage and implementation—the DNA). This information is translated into "executive mechanisms"—proteins—which are responsible for subsequent events (the emergence of new enzymatic activities, the formation of new structures using new structural components, the creation of new regulatory pathways through the synthesis of new signaling molecules, and so on). All of this is encoded within the program, namely the genome.

Since all the information regarding the structure of animal bodies is contained within their genes, biologists believed that the genomes of different species would vary significantly from one another. In reality, this is not the case: species that are completely distinct from each other possess many similar—even identical—genes. From an informational and hereditary perspective, all living organisms are incredibly unified. The morphological differences between species arise during evolution as a result of changes not in the genes themselves but in the "switches"—segments of DNA that regulate gene activity. If we want to understand what distinguishes one animal from another, we must focus not on genes but on the switches [39].

Previously, biophysicists believed that the development of an organism followed a rigid program encoded in the genes. Later on, they concluded that the programs embedded in DNA were secondary in nature. Supporters of the new hypothesis argue that genes continuously receive signals in the form of holograms—sort of "pictures" that show, step by step, the entire course of the organism's evolution. As a result, cells, organs, and the organism as a whole change in response to the changing information from the outside.

A.K. Korotkov from LITMO (now the St. Petersburg Institute of Precision Mechanics and Optics) has demonstrated the existence of a feedback mechanism: the organism continuously transmits information about its own state.

Scientists have learned how to use devices to extract information from those energy-information layers that do not disappear immediately after a person's death. Based on this information, they can determine how the person died. The researchers have also discovered microleptonic connections between people and their photographs.

Phenomenological studies of living systems make use of concepts from information theory. The existence and development of cells and organisms, which are self-regulating systems, require the generation and transmission of information through direct and feedback (reverse) communication channels.

The amount of information obtained when selecting P_1 messages out of P_0 equally probable messages is determined as follows:

$$I = k \log \frac{P_0}{P_1}, \quad (3.1)$$

where k is a numerical coefficient. If $P_1 = 1$, then $I = k\log P_0$.

For the quantitative calculations, it is necessary to determine the base of the logarithms and the value of 'k'. Using basis 2 and assuming k = 1, we have $I = \log_2(P_0/P_1)$ (3.1). This definition of the amount of information corresponds to problems involving a certain number n of different independent choices, each of which can be reduced to a binary alternative. Given that $R_0 = 2^n$, and according to (3.1), $I = \log_2 2^n = n$ bits.

The basic unit of information is a single bit, which represents one of two equally likely options (i.e., $P_0 = 2$). In physics, a bit corresponds to a physical system that can be switched into one of two distinct states, representing two logical values: "yes" or "no," "true" or "false," "0" or "1." At the atomic level, this can be illustrated by the two possible states of an electron in an atom, where "0" corresponds to the state with the lowest energy and "1" corresponds to the state with higher energy (an excited state). The definition of information used here is consistent with the binary number system, in which any number is represented as a power of 2 using the digits 0 and 1. Specifying the probability distribution for the bits implies that some prior information about the intended message is already available.

It is difficult to assess the amount of information contained within a cell or an organism. It can be argued that the information (i.e., the order and structure that enable biological functions) is entirely determined by the atomic structure.

G. Kastler estimated the amount of information required to select one atom from the full set of atoms in a cell and to specify its position with a precision determined by thermal fluctuations at body temperature [40]. According to his calculations, it takes 24.5 bits to represent one atom; approximately 10^{10} bits to represent an Escherichia coli cell, around 10^{15} bits to represent a mammalian zygote, and around 10^{25} bits to represent a human organism. The amount of information per amino acid is $I_a = \log_2 20 = 4.322$ bits, while the amount of information per nucleotide is $I_n = \log_2 4 = 2$ bits. Therefore, the minimum number of nucleotides required to represent one amino acid is $I_a/I_n = 2.161$. Information in living organisms is transmitted by molecules and ions and has a chemical nature.

The process of acquiring information essentially involves recognizing molecules through their interactions with other molecules; this recognition is determined by those molecular interactions. In this sense, an organism can be considered a chemical machine, which is what distinguishes it from "robots" created by human hands. In modern cybernetic devices, signals are of electrical or magnetic nature; in an organism, however, the signals are transmitted by molecules.

Recent studies have shown that each cell not only possesses a developmental program predetermined by DNA, but also has its own will—the will to resist destructive influences or to counter them in order to maintain its individuality. Until it becomes known how cells make these decisions, it will be impossible to control them.

A.A. Tyunyaev observes: "The world consists of Information about the interaction between various pieces of information. Information is a component part of the observed 'organism'; this 'organism' represents the smallest structural unit, and its structure, in the context of the specific task at hand, is considered as a unified whole." [41]

From the perspective of information theory, this means that the system generates information on its own, and the speed of this process is higher the greater the degree of chaos. Consequently, according to the theory of chaotic synchronization, it follows that the more intensively a system generates information, the harder it is to synchronize it—i.e., to make it behave in some other way. [42]

The person who will control the revolution in the field of information will, in essence, be exercising geopolitical control over the world, in one form or another.

G. Kastler introduced the signature principle into the consideration of molecular information [42]. The resulting genetic code remains unchanged; it serves as a carrier of this stable signature, which remains within the cell. The interaction between molecules involves the transmission of information. However, not all the information contained in the molecule is transmitted; only a portion of it is transferred, and this portion is referred to as the "signature."

Any information transmission is not perfect; during the process, some of the information is lost due to interference and errors. Such interference or errors are referred to as "noise."

The amount of information in a genetic message is equal to:

$$I \leq I_0 - Ish$$

I_0 is the initial (original) amount of information;

Ish: "Noise" (random errors) resulting from mistakes during the duplication of chromosomes in embryonic or somatic cells, as well as errors in the reception of genetic information, that is, during the synthesis of proteins.

The influence of interference cannot be completely eliminated, but every possible measure is always taken to identify inaccurate information. The credibility of the information is assessed accordingly. The utilization rate of that information, taking into account the parameters of the samples as well as the number of measurements conducted, plays a significant role.

3.2. THE CONCEPT OF SYSTEM COMPLEXITY AND THE FOUNDATIONS OF A SYSTEMATIC APPROACH IN THE PROCESS OF STUDYING OBJECTS

One of the earliest definitions of a complex system was given by N. P. Bouslenko in 1973. He stated that a system should be considered complex if it consists of a large number of interconnected and interacting elements, and that a complex system is capable of performing a complex function. Later, in his work [43], he provided a definition that characterized some of the fundamental properties of such systems: "A complex system is a multi-level structure composed of interacting elements, which are organized into subsystems of various levels." The mathematical model of a complex system "consists of the mathematical models of the individual elements as well as the mathematical model of the interactions between those elements."

As examples of complex systems, one can mention energy complexes, telephone networks in large cities, information systems, industry-specific automated control systems, and undoubtedly the structures of living organisms. Living organisms belong to the category of complex natural systems. They continuously receive information from the external environment in various forms and process it in such a way that necessary and highly diverse reactions occur within the organism. Within the cells of an organism, millions of chemical reactions take place every minute; depending on the concentration of certain regulatory proteins, genes responsible for producing the desired proteins [44] are switched on or off.

Systemic approach: A methodological perspective in science whose primary goal is to develop research methods for studying complex, organized systems of various types and classes. This approach is most widely applied in the study of complex, evolving entities—hierarchical, typically self-organizing biological systems, etc. Within the systemic approach, significant emphasis is placed on identifying the probabilistic nature of the behavior of the systems under investigation.

In their search for methods to model complex systems, researchers turn to various branches of mathematics, such as the theory of fuzzy sets and tools for interactive computer graphics. The guiding principle for their approach is Gödel's incompleteness theorem, which demonstrates that it is impossible to completely formalize all knowledge. This necessitates the use of a combination of both mathematical and heuristic methods in conducting systematic analyses [7].

Natural systems possess a particular characteristic: they do not operate according to absolute laws. Deviations from the observed patterns are common in natural systems, which is inherent in living entities. Nature reveals to us not merely a higher degree, but an entirely different level of complexity. It turns out that nearly all natural structures exhibit fractal properties. For instance, the chaotic nature of the heartbeat allows it to respond flexibly to changes in physical and emotional demands, adapting accordingly. However, it is also known that a regular heartbeat can lead to fatal consequences over time. One reason is that the heart may lack the "mechanical strength" to compensate for external influences. Regularity indicates a reduced resistance of the organism to random environmental disturbances; at that point, the system is no longer able to adequately track changes and respond to them flexibly enough. Clearly, any complex system functioning in a dynamic environment must possess such adaptability and controllability. This is essential for its survival and successful evolution (see "Science and Life," 2001, No. 5).

An analysis of the definitions of the concept "system" allows us to broadly consider a system (S) as an ordered pair of sets: a set of corresponding elements (A) and a set of relationships (R) between the elements of set A.

S = (A, R).

Any binary relation that possesses properties such as reflexivity, symmetry, and transitivity is commonly referred to as an equivalence relation (e.g., identity, equality, similarity, proportionality, etc.). As for the importance of these relations, it's sufficient to say that any classification is based on one or more of them.

3.3 CLASSIFICATION OF RECOGNITION SYSTEMS

Today, classifications have become one of the main methods and products of research, as they enable the establishment of connections between objects, phenomena, and processes at different levels.

Recognition is a scientific and technical field that involves the development of methods and systems for determining to which of pre-defined classes an object belongs.

The recognition process is based on comparing the features of the object under study with the features of other known objects. By doing so, a conclusion about the most likely correspondence between them is drawn.

Recognition methods are used in technical and medical diagnostics, forensic science, and other fields. [46].

Depending on the classification principle and the nature of the information, recognition systems can be divided into deterministic, probabilistic, logical, structural, and combined types.

3.4. DETERMINISTIC SYSTEMS

The principle of determinism states that there is a strict, one-to-one correspondence between the quantities that characterize the state of a system at a given moment in time and the values of those quantities at any subsequent (or previous) moment in time.

In the 18th century, P. Laplace successfully applied the principle of determinism to celestial mechanics, studying the motion of the planets in the solar system. He believed that the principle of determinism applied to all phenomena in nature, and that by knowing precisely the state of the universe at any given moment, one could unambiguously predict its future. [47] As an example, we can consider Kepler's discovery in the 17th century of one of the laws that bear his name. By comparing the powers of the distances with the times of celestial revolutions, Kepler found that the squares of these periods are related to each other as the cubes of the major axes of the orbits.

Later on, Newton discovered the law of universal gravitation: the force of mutual attraction between two material particles (whose sizes are so small that they can be neglected) is proportional to the product of their masses and inversely proportional to the square of the distance between them. This is an example of a functional relationship between the four quantities involved in the law; knowing the values of three of these quantities allows for the determination of the value of the fourth one.

In the 20th century, relying on Newton's laws, scientists were able to determine the mass of molecules and atoms, and later they were able to accurately calculate the trajectories of spacecraft.

The geneticist A.A. Malinovsky, who was also interested in mathematics and statistics, conducted a statistical analysis of the relationship between theoretical thinking and what might seem like a factor far removed from science—morality. Malinovsky found that only individuals with outstanding moral qualities also became outstanding theorists. It is clear that there is no genetic or physiological basis to support this statistical conclusion, but a sociological basis is entirely possible. Truth is harder to obtain than gold, but the important thing is to realize that whereas gold can be obtained by any means, truth cannot be obtained by just any means. [48].

A deterministic model is a model that assumes the behavior of the system being simulated to be definite and predictable under given conditions. The parameters and variables of a deterministic model are fixed, non-random values. To build such a model, it is necessary to understand the behavior and interactions of all components of the system being modeled, which is only possible in relatively simple situations. In the absence of uncertainty, the clarity and simplicity of a deterministic model make it attractive for analysis; it can provide sufficient information for decision-making purposes.

In real-world technical, biological, economic, social, and other complex systems, there is a significant amount of uncertainty in the input data. This makes it practically impossible to accurately predict the behavior of these systems using deterministic models. As a result, researchers much more frequently resort to stochastic models [47].

3.5. PROBABILISTIC SYSTEMS

It is generally believed that the theory of probability originated from the correspondence between two great scientists: Blaise Pascal (1623–1662) and Pierre Fermat (1601–1665).

A significant contribution to the theory of probability was made by the French physicist and mathematician Pierre-Simon Laplace (1749–1827). He approached the concept of belief in randomness with caution, stating: "The notion of randomness merely signifies our ignorance of the causes of the events we observe, as well as of the reasons why these events occur in a sequence that appears to us random." Another notable contributor was Galileo Galilei (1564–1642), the renowned scientist and naturalist. His work "On the Outcome of Dice Games" focused on calculating the number of possible outcomes when three dice are rolled; each die had a specific number (first, second, third). Thus, the dice game represents one of the simplest examples of a random process. The die, with its cubic shape (a three-dimensional representation of a square), symbolizes the final stage of a cyclical motion and, therefore, figuratively fits into a circle. The cubic form embodies stability, static perfection, and truth—truth that remains constant throughout time and space, from all perspectives. The sides of the cube can also be seen as representing the sacred cycles of existence. Hence, rolling a die serves to reveal aspects of combinatorics that have not yet been manifested;

these combinatorial possibilities are described by the die's structure as if they had already occurred, and thus remain unchanging. [39]

However, in Galileo's reasoning, the focus is not on the probabilities of random events, but rather on the numbers of favorable outcomes. More details about these chances (odds) were provided by Huygens in his works. Later on, with his proposals, Huygens introduced the concept of mathematical expectation for a random variable that can assume two or three values.

There are two definitions of probability: the classical definition and the statistical definition. Both of these definitions were first clearly formulated by J. Bernoulli.

3.5.1. Basic definitions and concepts of probability theory

Probability theory is a branch of mathematics in which, based on the probabilities of certain random events, the probabilities of other events that are related to those first events in some way are determined.

Probability is a mathematical numerical characteristic that represents the likelihood of the occurrence of some random event A under certain specified conditions, which may be repeated an unlimited number of times. The probability of event A, denoted as P(A), is defined as the ratio of the number m of elementary events that are favorable to event A to the total number n of all possible elementary events:

$$P(A) = \frac{m}{n}. \quad (3.2)$$

A random event is an event that may occur or not occur when certain conditions are met (i.e., during a particular experiment), and for which there is a specific probability of occurrence. The fact that a random event has a certain probability P (where $0 \le P \le 1$) is reflected in the following phenomenon: when the number of experiments is large, the frequency of the event occurring approaches P.

Independent Events: A set of events is called mutually independent (or simply independent) if, in addition to their pairwise independence, any one of them is independent of the product of any number of the others. If events A and B are independent, then the following holds:

$$P(AB) = P(A) \cdot P(B). \quad (3.3)$$

Dependent events: Two events are considered dependent if the probability of one of them occurring depends on the occurrence of the other.

An event A is called a *certain event* if P(A) = 1.

The opposite of an event is an event that…

The event "A" does not occur, and it is represented by the symbol "Ā".

The conditional probability P(A|B) of event A given event B is the probability of event A occurring under the condition that event B has already occurred. In this case, P(A|B) will be equal to:

$$P(A|B) = \frac{P(AB)}{P(B)}. \quad (3.4)$$

The Multiplication Theorem

The probability that two events occur simultaneously is equal to the product of the probability of one of the events occurring multiplied by the conditional probability of the other event occurring, given that the first event has already taken place. In other words:

$$P(AB) = P(A)\,P(B|A) = P(B)\,P(A|B). \quad (3.5)$$

The formula for the full probability...

Event B may occur if either of the events A_1, A_2, ... A_n takes place. This means that the occurrence of event B entails the occurrence of one of the events A_1B, A_2B, ..., AnB, no matter which. Therefore, the event B = A1B + A2B + ... + A_nB. Since events A_1, A_2, ... A_n are pairwise incompatible by condition, it is obvious that events A_1B, A_2B, ..., AnB are also not compatible with each other. Applying the addition theorem for mutually exclusive events, we obtain:

$$P(B) = P(A_1B) + P(A_2B) + ... + P(A_nB). \quad (3.6)$$

By applying the theorem of probability multiplication further, we will obtain the formula for the total probability.

$$P(B) = \sum_{i=1}^{n} P(A_i)P(B|A_i)\,. \quad (3.7)$$

The application of probabilistic recognition methods assumes the existence of probabilistic relationships between the features of the recognized objects and the classes to which those objects belong.

There are many well-known examples in history where scientists' ideas were used in later periods and helped shape the development trends of various fields of knowledge. A similar situation occurs with the work of Thomas Bayes on probability theory. The results of his research were published at the end of the 18th century, but the practical application of Bayesian methods did not begin until the 1990s, coinciding with the computer revolution.

The Bayes' formula is the foundation of the inductive approach to recognition. It should be noted that if one of the object classes is missing from the training dataset, any recognition procedure will perform poorly. A key aspect of justifying inductive recognition procedures is the presence of all classes in the dataset, as well as averaging the error rates across multiple training datasets. Therefore, attempting to evaluate inductive recognition procedures without taking these factors into account will not yield positive results.

The Bayesian procedure has several notable characteristics: it is optimal in terms of performance, its error estimates are polynomial in relation to the complexity of the input data, and its error rate decreases rapidly as the size of the training datasets increases. There is every reason to believe that inductive procedures play a crucial role in the mechanisms responsible for modifying and altering the functional properties of genes.

3.5.2. Historical prerequisites of the Bayesian method

Bayes' fundamental research in the field of probability theory was presented in his work "An Essay towards Solving a Problem in the Doctrine of Chances." This paper was published in 1764 in the "Philosophical Transactions of the Royal Society of London," where his world-renowned fame began [49].

The further development of the Bayesian method is owed to Laplace. It was Laplace who laid the groundwork for the subjective interpretation of prior probabilities. In the mid-19th century, Boole and Venn pointed out the arbitrary nature of using the Bayesian method to obtain probabilities that cannot be verified in any way. It was during this period that the debate between proponents and opponents of the Bayesian method began and continues to this day. This debate dies down and flares up again from time to time, depending on new arguments put forward by both sides. Some scientists believe that Bayesian theory has reached a "dead end"—a situation stemming from the lack of a sufficient justification for the rules used in selecting prior distributions. According to Neyman, the way out of this dead end lies in empirical Bayesian estimation, which is based on approximating the prior distribution using observational data. D. Lindley proposed a "correction" of empirical Bayesian estimation by introducing a random hyperparameter of the distribution of a random variable into the family of prior distributions [50]. Lindley stated: "The developments in the field of mathematical statistics will continue in the 21st century." [51]

In his work [52, 53], V.V. Nalimov proposed the hypothesis that the Bayesian model plays a primary role in natural mechanisms as well as in many other aspects of human activity, particularly in the probabilistic language model based on that framework. V.V. Nalimov believes that this model "explains and facilitates an understanding of various phenomena: the interpretation of texts that constantly contain words with ambiguous meanings; the creative process of generating new texts; human behavior and how it leads to changes in one's value systems in new situations; the semantic complexity of the individual; the concept of self-organization as a creative process that occurs in a certain unified manner within nature; and aspects of biological evolution and human creativity." V.V. Nalimov also put forward a strategic perspective that underlies the idea of the genome's "language-like" structure. He viewed the universe as a rational linguistic entity and, following Spinoza, identified Nature with the Creator God.

3.5.3. General principles of the Bayesian methodology

The Bayes' formula is a consequence of the theorem of multiplication and the formula for total probability. Let's recall that event B, whose probability we need to determine, can only occur if one of the events A_1, A_2, …, A_n occurs; these events form a complete set. In other words, suppose…

$$B = \sum_{i=1}^{n} BA_i \,, \tag{3.8}$$

Events BAi and BAj with different indices i and j are not compatible with each other. According to the theorem of probability addition…

We have:

$$P(B) = \sum_{i=1}^{n} P(BA_i) \,. \tag{3.9}$$

Let's see how the probabilities of the hypotheses will change after this. In other words, we'll find the conditional probability for each hypothesis. According to the theorem of multiplication, we have: $P(A_iB) = P(B)\ P(A_i|B) = P(A_i) \cdot P(B|A_i)$, from which

$$P(A_i|B) = \frac{P(A_i)P(B|A_i)}{P(B)}, \tag{3.10}$$

$$P(A_i|B) = \frac{P(A_i)P(B|A_i)}{\sum_{i=1}^{n} P(A_i)P(B|A_i)} \,. \tag{3.11}$$

This formula is known as Bayes' Formula.

The Bayes' formula is applied when an event B occurs, which can only happen under one of the hypotheses A_1, A_2, …, A_n (which together form a complete set of possible events). it is necessary to quantitatively re-evaluate the prior probabilities of these hypotheses $P(A_1)$, $P(A_2)$, …, $P(A_n)$, known before the experiment was conducted. Specifically, we want to determine the posterior probabilities (obtained after the experiment) of the hypotheses $P(A_1|B)$, $P(A_2|B)$, …, $P(A_n|B)$.

The significance of Bayes' formula lies in the fact that, as we obtain new information, we can re-evaluate and adjust the hypotheses proposed before the experiment (the decisions made or the assumptions of the model) based on the transition from their prior probabilities to their posterior probabilities.

This approach, known as Bayesian methodology, enables the adjustment of management decisions, the estimation of unknown parameters in the distribution of the features being studied in statistical analysis, and so on. [54].

We will understand the methodology of the Bayesian approach in the commonly accepted meaning of this term, that is, as a set of principles for constructing models and methods for conducting research.

First of all, it is necessary to clarify which principles constitute the foundation of the Bayesian approach. Often, when referring to Bayesian theory, people mean a particular aspect of it, such as the subjective interpretation of probability, a methodology based on Bayes' theorem, or sampling-based Bayesian decision theory, etc. However, in its modern understanding, the Bayesian approach is a unified theory that seamlessly integrates the following three main principles:

Position 1

The parameter of the system or model being studied is random, and this randomness can be understood not only in the common sense of the word but also as uncertainty. A random parameter is assigned a prior distribution. The interpretation of judgments in Bayesian methodology is always probabilistic in nature. Probabilistic judgments can be represented in one of three ways:

- This frequency/objective interpretation of probability is extremely rare, as it requires a large number of past experiences to be valid.
- Rational degrees of certainty, which for practical purposes essentially boil down to the mathematical expression indicating the absence of any prior knowledge.
- These subjective levels of trust reflect the researcher's personal attitude towards the phenomenon or system being studied.

The areas of application for these methods hardly overlap with each other.

In the first case, when there is a large amount of past experimental data, both from rationalist and subjective perspectives, the levels of confidence will inevitably coincide with the relative frequencies. In the absence of any knowledge at all, the subjective levels of confidence should coincide with the rational levels, which means that an equal distribution must be assumed. In all other cases (which constitute the vast majority), the subjective levels of confidence must be adopted as the basis for decision-making.

Position 2

The results of observations and the prior distributions are combined using Bayes' theorem to obtain the posterior distribution of the parameter. This process can be thought of as a gradual accumulation of information. At the beginning of the study of a phenomenon, a researcher with certain qualifications and experience from previous work has some understanding of the properties of the subject under investigation. This understanding includes both informal insights and empirical data obtained from similar studies conducted in the past. As the subject is tested, new information emerges in the form of additional empirical data, which modifies the initial perception (i.e., the probabilistic judgment) of its properties. Consequently, the prior understanding is gradually revised and re-evaluated. At any given point in time, we can provide a complete answer regarding the properties of the subject, and this answer will be exhaustive in the sense that we have utilized all available information.

If two researchers start with different prior distributions (perhaps due to different initial information), their posterior distributions will converge over time. As a result, using Bayes' formula, we can more accurately recalculate the probability by taking into account both the previously known information and the data from the new observations.

Position 3

The statistical conclusion or decision rule is derived based on the maximum expected utility, particularly with the aim of minimizing the losses associated with the application of that rule. The ultimate outcome of the aforementioned Bayesian procedures is a posterior distribution of the parameter that characterizes the main properties of the system or phenomenon under study. This distribution provides a clear and comprehensive understanding of the uncertainty surrounding the parameter's value. However, in real-world practical situations, a more concise decision rule is needed, one that allows the parameter to be expressed as one or several numerical constants serving as estimates of the unknown parameter [55].

Bayes' Theorem provides the methodological foundation for the process of transitioning from prior information, which is formalized as a prior distribution, to posterior information by incorporating empirical data.

The diagram presented by Zellner in Figure 3.1 clearly illustrates the process of updating probabilities upon the receipt of new data. As more sample information is accumulated, this sample-based distribution begins to dominate the posterior probability distribution. The density of the posterior distribution increasingly concentrates around the true value of the parameter [56].

The scheme presented below is called the orthodox Bayesian procedure. Its difference from later modifications lies in the fact that, as empirical data is collected, the prior distribution remains unchanged.

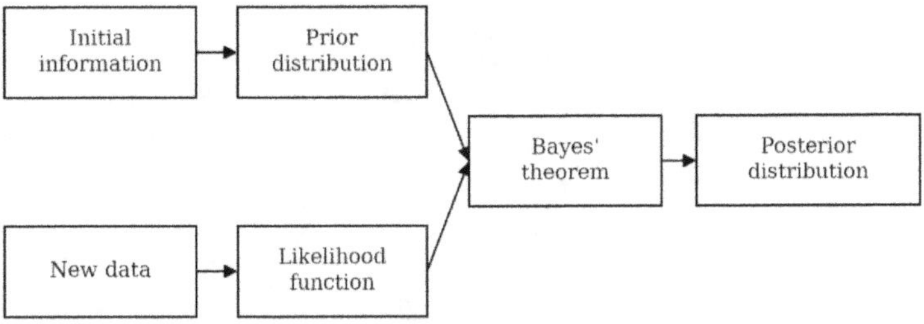

Figure 3.1. Schematic representation of the orthodox Bayesian procedure

Leamer proposes a more complex modification to the orthodox framework, one that allows for the adjustment of prior confidence levels. The complete Leamer framework is depicted in Figure 3.2. [51].

Leamer's approach is based on the idea that individuals are born with a countless number of subconscious judgments, each associated with a certain inherent level of confidence. Experience leads to the selection of a relatively small subset of these judgments from which conscious beliefs are formed; these conscious beliefs are then supplemented by additional notions acquired through new experiences. Based on this collection of judgments, a set of working hypotheses and their corresponding levels of confidence is established. In practical situations, these confidence levels represent rough approximations of the levels that a researcher with unlimited memory and similar cognitive abilities would have, according to Bayesian theory. The theoretical part of the process involves determining these initial confidence levels for the hypotheses. Subsequently, these confidence levels are revised using Bayes' theorem by incorporating empirical data to obtain more accurate posterior estimates.

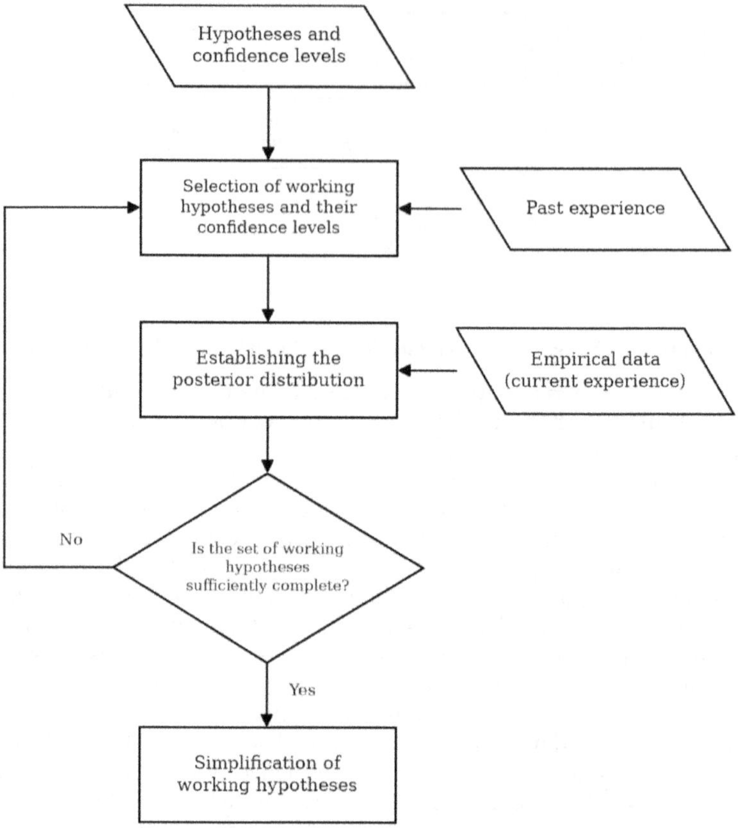

Figure 3.2. The complete, orthodox Leamer diagram, which includes the possibility of adjusting the initial levels of trust

The characteristics discovered in the data may prompt the researcher to re-evaluate many of the decisions made earlier. The data might suggest that, for example, one of the rejected hypotheses should be included among the viable options, or that a completely new hypothesis should be considered. By revising one or more of their previous decisions, the researcher will re-analyze the same dataset using the new hypotheses or prior distributions.

Evans goes even further. He argues that Bayesians are always willing to update their prior probabilities in the light of new observations. As a result, when the new prior distributions are combined with the quantitative data, both the Bayesian posterior distribution and the subjective prior distribution become consistent with each other. The greatest effect is achieved when there are multiple prior distributions, and the researcher selects the one that is most consistent with the empirical evidence. Evans believes that this approach could potentially eliminate subjective probabilities altogether from the Bayesian framework. However, this last claim is rather contentious. The modifications discussed pertain to the rational Bayesian procedures [51].

The main characteristic of the described Bayesian procedures (both the orthodox and the rational ones) is that they can always be used for specific calculations and are applicable to the analysis of a wide range of problems in the field of reliability theory and other areas of science. The Bayesian approach is universal and possesses the property of internal unity.

3.6. LOGICAL SYSTEMS

The thinker who laid the foundations for logic and is considered the "father of logic" is Aristotle (384–322 BCE). Aristotelian syllogistic theory is his major achievement in the field of logic; it constitutes a system based on axiomatic deduction.

The paths of mathematics and logic diverged, and for centuries, they were considered entirely different sciences: logic was generally classified as a humanities discipline, which is the very opposite of the exact sciences, whose core is mathematics. Later on, logic made use of mathematical methods; thereafter, it underwent a sort of "rebirth" and rapidly moved towards reuniting with mathematical knowledge. Modern mathematical logic draws upon the works of J. Boole and A. De Morgan, in which logical science acquired its symbolic language and the rules for manipulating those expressions. The concept of Boolean algebra is already familiar to many non-mathematicians and non-logicians alike, and the notion of a Boolean variable has become commonplace among programmers, operators, and anyone who works with computers.

In logical systems used for designing recognition algorithms, logical recognition methods based on discrete analysis are employed. In general, the application of such methods requires the presence of logical relationships expressed through a system of Boolean equations. In these equations, the variables represent the logical features of the objects to be recognized, while the unknown quantities denote the classes to which those objects belong. Logical computations and logical programming have recently received increasing attention, as they enable a significant expansion of the range of problems that can be solved by computers.

In practical situations, the information available about features is often limited. Based on this limited data, it is necessary to construct an overall understanding of the class in question and identify the patterns inherent in its characteristics. A pattern is generally understood as a certain relationship between the features of observed phenomena—such a relationship that is strong enough to be detected even when observing only individual, randomly selected objects from the class being studied. For this to be possible, the sample must be representative: the objects are chosen independently of each other and in sufficient quantity.

The weakest connection is defined by a single element in the set of prohibitions—namely, the specific element that is prohibited. In general cases, a connection encompasses multiple features; the weakest type of connection covers all features. Suppose r features are related (denoted as $r \leq n$), and this relationship represents a prohibition on some combination of their values. Let's represent this relationship not as a binary vector but as a ternary vector, whose components take values from the set {0, 1, –}. The r components corresponding to the related features will assume the value 0 or 1, while the remaining n–r components will have the value "–", which symbolizes uncertainty. Such a connection is called an "implicative connection of rank r". An implication is a formalization of the "if..., then..." type of relationship commonly found in natural language expressions.

There is a problem in the process of constructing knowledge that makes it difficult to use traditional mathematical tools. This problem relates to the description of concepts that operate with qualitative characteristics of objects (such as "many," "few"; "high," "low"; "far away," "very close," and so on).

It should be noted that, according to the principle of consistency (also known as the Law of Excluded Middle), among two statements: A and non-A, one must be true and the other must be false; there cannot be a third possibility. However, everything in the world is at least three-dimensional in nature: there are "yes" and "no", but also the ambiguous option of "maybe".

As mentioned above, the relationship formed by a triple vector containing components from the set {0, 1, "–"} (where "–" represents uncertainty) is called an implicative relationship. It is natural to expect that implicative relationships are stronger the fewer features they connect [57].

In ternary code, unlike binary code, there is no need to distinguish between "unsigned numbers" and "signed numbers". This reduces the number of conditional instructions required by a factor of two and significantly simplifies their use. The idea of ternary code has been advocated for many years by Nikolai Petrovich Brusencov, a scientist from the Department of Computational Mathematics and Cybernetics at Moscow State University (MSU). This idea deserves to be implemented [58].

In the work by N.P. Brusencov, [59], the main advantages of the ternary code system are discussed, the efficiency of three-digit logic is demonstrated, and the feasibility of economical and reliable transmission and conversion of three-digit signals is explored.

In problems solved by intelligent systems, it is often necessary to use incomplete or imprecise information that cannot be interpreted as either completely true or completely false (i.e., not represented by 0 or 1). There are values whose certainty is expressed by a numerical intermediate value, such as 0.6. To address these issues, in the early 1970s, the American mathematician Lotfi Zadeh proposed a formal framework for fuzzy algebra and fuzzy logic.

Systems with fuzzy logic are suitable for explaining the conclusions derived from them, but they cannot automatically acquire knowledge for use in the inference mechanisms. Moreover, the main drawbacks of fuzzy-logic systems are related to the fact that the initial set of fuzzy rules is formulated by an expert—a human—and may turn out to be incomplete or contradictory.

3.7. STRUCTURAL SYSTEMS

In these systems, special grammars are used to construct recognition algorithms. These grammars generate languages consisting of sentences, each of which describes objects belonging to a specific class. To apply structural recognition methods, it is necessary to have sets of sentences that describe all the objects belonging to all the classes in the recognition system's class hierarchy. The set of sentences must be divided into subsets based on the number of classes in the system. The elements of each subset are sentences that describe objects belonging to that particular subset. Thus, the a priori descriptions of the classes are represented by sets of sentences, with each sentence corresponding to an object belonging to a specific class.

3.8. COMBINED SYSTEMS

In these systems, specially developed methods for calculating evaluations are used to construct recognition algorithms. Such recognition algorithms are referred to as "evaluation calculation algorithms." Their implementation requires the use of tables that contain objects belonging to respective classes, as well as the feature values that characterize those objects. The features can be either deterministic, logical, probabilistic, or structural in nature.

3.9. INTELLIGENT SYSTEMS

An intelligent system is a technical or software-based system capable of solving problems that are traditionally considered to be creative in nature and belong to a specific domain of knowledge. The knowledge related to that domain is stored within the memory of the system (see the "Explanatory Dictionary of Artificial Intelligence").

The source of the disturbing influence on an intelligent system is its environment, and the function of the system should ultimately be to compensate for this influence. In the early 1990s, a new, comprehensive discipline began to emerge; today, it is recognized as an independent field of study known as Computational Intelligence (CI). CI works with "soft computing" methods that are based on approximate models, including techniques for approximate reasoning, as well as computational methods based on functional approximation, random search, and optimization.

Artificial Intelligence (AI) deals with rigorous computations based on precise models that incorporate reasoning grounded in symbolic logic, as well as classical methods for computing and information retrieval.

3.9.1 Classification of Intelligent Systems

Intelligent systems that supplement or replace human information-processing capabilities when dealing with incomplete or contradictory information can be roughly divided into two main categories, based on the approach they use to perform these supplementary or replacement functions:

1) These systems are designed to take into account how humans (as objects) perceive their external environment.
2) Systems that focus on understanding a person's inner world.

The systems within each major class can be divided, based on their properties, into primarily informational (passive) and control-oriented (active) types.

Intelligent systems are most often used to solve complex problems, where the main difficulty lies in the use of poorly formalized knowledge from expert practitioners, and where logical information processing prevails over computational information processing.

3.10 EXPERT SYSTEMS AND METHODS FOR EXPERT EVALUATION

Expert systems can be classified as general-purpose artificial intelligence systems.

The use of experts' knowledge helps to obtain and refine descriptions of problematic situations. To convert a high-quality descriptive account into a formal specification, methods based on the intuition and experience of specialists are employed. The collective decision is made through a deterministic procedure. Currently, techniques such as "brainstorming" and "expert evaluations" are widely used as such procedures.

The concept of a "brainstorm" has gained widespread popularity since the early 1950s as a method for stimulating a group of people to quickly generate a large number of ideas.

The "brainstorming" method can be applied to any problem as long as it is presented in a simple and clear manner. This approach can be used either when the problem has not yet been fully defined or later on, once more complex aspects have been identified. It can also be utilized for generating information, not just ideas [60].

When considering methods for expert evaluation, what is often meant is a set of logical and mathematical techniques and procedures aimed at obtaining information from experts that is necessary for preparing and selecting rational decisions.

The basis for the application of all expert evaluation methods lies in the expert's practical knowledge and their principles related to the practical activities outlined in [61]. This enables experts to determine the importance and significance of certain issues, the potential prospects of a particular research direction, as well as the timing of various events. The expert evaluation methods used can be divided into two categories: individual expert evaluation methods and collective expert evaluation methods, which can further be subdivided into several groups. Individual expert evaluation methods include techniques such as interviews and analytical assessments.

Methods for collective expert evaluation include: the committee method, the relative evaluation method, the Delphi method, and others. [62].

Unlike the traditional approach to reaching consensus among experts through open discussion, the "Delphi method" (named after the ancient Greek city of Delphi) involves a complete rejection of group discussions. This is done to minimize the influence of psychological factors such as the tendency to align with the opinions of the most authoritative specialists or to follow the majority. In the Delphi method, direct debates are replaced by a carefully designed program of sequential individual surveys, usually conducted in the form of questionnaires. The anonymity of the survey is one of the main factors contributing to the popularity and effectiveness of this method.

The experts' findings are summarized and, along with new additional information, made available to them. They then revise their initial responses accordingly. After the expert evaluations have been compiled and their consistency analyzed in order to enhance the credibility of the expert opinions, a further round of evaluation may be conducted if necessary. In this case, the experts are informed about the results of the previous round's consensus-based evaluations, and they provide feedback by adjusting their own judgments. The criterion used to determine the consistency of expert opinions is typically a measure of agreement, expressed as the ratio of the number of consistent evaluations that fall within a confidence interval to the total number of evaluations in the expert group.

Many studies have been dedicated to examining the characteristics and potential applications of expert evaluations; these studies consider various aspects, including:

1) Issues related to the formation of expert groups (including requirements for experts, group sizes, and matters concerning the training of experts to assess their competence).
2) Forms of expert survey (various types of questionnaires, interviews, and mixed formats) and methods for organizing the survey;
3) Approaches to evaluating experts (ranking, normalization, various types of sorting, including methods for pairwise comparisons, etc.);
4) Methods for processing expert evaluations;
5) Methods for determining the consistency of experts' opinions (the reliability of expert evaluations, as well as techniques for enhancing the consistency of those evaluations through appropriate processing of the results of expert surveys) [7].

Expert systems are only effective in specific "expert" domains where the empirical experience of specialists is relevant.

A detailed review of the forms, methods for obtaining, and processing expert evaluations can be found in [60, 63].

The difficulties involved in conducting ranking studies to determine preferences when there are a large number of factors can be alleviated to some extent if experts are asked to compare these factors in pairs, with the goal of identifying the most important factor in each pair. In the paired comparison method, the expert compares the options in pairs and then selects one of the two. Typically, when comparing two options, the evaluation is limited to a simple determination of which one is preferable to the other.

The problem of selecting experts has two aspects: choosing an individual expert and choosing a group of experts. The selection process must ensure that for each issue, at least one expert's opinion can be obtained, while minimizing the time or resources required to conduct the expert consultation. An important and responsible step in the expert evaluation process is the final stage of analyzing and processing the results of the survey.

3.10.1. Example of using the expert evaluation method

When dealing with forecasting tasks, there is often a lack of sufficient statistical data. This deficiency can be partially compensated for by information based on assumptions and judgments made by experts and specialists. The use of such information is particularly fruitful when specialized logical techniques and mathematical methods (known as expert evaluation methods) are applied to its collection, processing, and analysis. Experience, intuition, and knowledge enable experts to assess many issues arising in situations of risk and uncertainty, as well as to evaluate relevant factors. Intuitive foresight forms the basis for expert evaluations.

It is of great importance to verify the consistency and reliability of the expert evaluations. Experts must rank the factors, that is, arrange them in order according to the increasing or decreasing degree of their influence on the final quality indicators of the system: $x_1, x_2, \ldots, x_n$. The results of the experts' ranking of the factors m can be presented in Table 3.1. The columns of this table represent the ordinal numbers in the ranking sequence specified by the j-th factor from expert i. The elements in the table are called "ranks". In real-world scenarios, experts' opinions will differ at least with regard to some factors; in other words, different numbers will appear in several columns of the matrix. The task at hand is to assess the degree of consensus among experts regarding the rankings of these factors.

Table 3.1

E|f	X_1	X_2	…	X_j	…	X_n
1	X_{11}	X_{12}	…	X_{1j}	…	X_{1n}
2	X_{21}	X_{22}	…	X_{2j}	…	X_{2n}
…	…	…	…	…	…	…
i	X_{i1}	X_{i2}	…	X_{ij}	…	X_{in}
…	…	…	…	…	…	…
m	X_{m1}	X_{m^2}	…	X_{mj}	…	X_{mn}

To achieve this, certain actions need to be performed on the elements of the resulting table.

Calculate the sum of the ranks for each factor, that is, the sums of the values in the corresponding columns.

$$\sum_{i=1}^{m} x_{i1}, \sum_{i=1}^{m} x_{i2}, \dots, \sum_{i=1}^{m} x_{in}. \tag{3.12}$$

The average value of this series will be equal to:

$$\bar{x} = \frac{1}{n}\sum_{j=1}^{n}\sum_{i=1}^{m} x_{ij} = \frac{1}{n}\sum_{i=1}^{m}\sum_{j=1}^{n} x_{ij} \tag{3.13}$$

$$\sum_{j=1}^{n} x_{ij} = \frac{n(n+1)}{2} \tag{3.14}$$

The result will be the sum of the numbers in the natural number sequence from 1 to n.

Then, the expression (3.13) will be:

$$\bar{x} = \frac{1}{n}\sum_{i=1}^{m}\frac{n(n+1)}{2} = \frac{m(n+1)}{2} \tag{3.15}$$

The mean value of the series (3.12) is…

We will subtract the average value (3.15) from each number in the sequence (3.12), square each resulting difference, and then sum all n obtained differences squared (d_j).

$$\sum_{j=1}^{n} {d_j}^2 = \sum_{j=1}^{n}[\sum_{i=1}^{m} x_{ij} - \frac{m(n+1)}{2}]^2 \tag{3.16}$$

Expression (3.16) will be maximized when all m experts rank the factors in the same manner. To assess the degree of agreement among the ranking sequences of all experts, the rank correlation coefficient (agreement coefficient) can be used.

$$W = \frac{\sum_{j=1}^{n} {d_j}^2}{max(\sum_{j=1}^{n} {d_j}^2)}. \tag{3.17}$$

The numerator represents the value of the quantity calculated according to (3.16) based on the results of expert surveys, while the denominator represents the maximum possible value of that quantity.

The value of W has two extreme values:

W = 0 – if there is no consensus among experts regarding the ranking.

W = 1 – if there is complete agreement among the experts.

The denominator (3.17) is derived from the condition that all factors are ranked identically by every expert.

Let's consider the case where the ranks form a natural number sequence. In that case, the rank matrix will have the appearance shown in Table 3.2.

Table 3.2

E|f	X_1	X_2	....	X_n
1	1	2	...	n
2	1	2	...	n
...	...	...	...	...
m	1	2	...	...

The values in the columns representing rank sums correspond to the series: m, 2m, …, nm.

We will calculate the average value (3.15) for each element in this sequence.

$$m - \frac{1}{2}m(n + 1), 2m - \frac{1}{2}m(n + 1), \dots, nm - \frac{1}{2}m(n + 1). \qquad (3.18)$$

We will then transform the elements of this sequence.

$$\frac{1}{2}m(1 - n), \frac{1}{2}m(3 - n), \dots, \frac{1}{2}m(n - 1). \qquad (3.19)$$

Now, square each of the terms and then sum them together:

$$max(\sum_{j=1}^{n} d_j{}^2) = \frac{1}{4}m^2[(1 - n)^2 + (3 - n)^2 + \dots + (n - 1)^2] =$$

$$= \frac{1}{4}m^2(1 - 2n + n^2 + 9 - 6n + n^2 + \dots + n^2 - 2n + 1). \qquad (3.20)$$

The free terms represent the sum of the odd-numbered elements of a natural number sequence; therefore:

$$\textstyle\sum_{j=1}^{n} x_{jn} = \frac{4n^3 - n}{3}; \qquad (3.21)$$

Then: $n^2 = \frac{1}{4}m^2(nn^2 - 2nn^2 + \frac{4n^3 - n}{3}) =$

$$= \frac{1}{2}m^2(3n^3 - 6n^3 + 4n^3 - n) = \frac{1}{12}m^2(n^3 - n), \qquad (3.22)$$

$$W = \frac{\sum_{j=1}^{n} d_j{}^2}{\frac{1}{12}m^2(n^3 - n)}. \qquad (3.23)$$

If an expert is unable to establish a hierarchical difference between several factors and assigns them all the same rank, then the calculation of the concordance coefficient is performed using the following formula:

$$W = \frac{\sum_{j=1}^{n} d_j^2}{\frac{1}{12}m^2(n^3 - n) - m\sum_{i=1}^{m} T_i}, \tag{3.24}$$

$$where\ T_i = \frac{1}{12}\Sigma(t_i^3 - t_i);$$

t_i – represents the number of occurrences of a certain rank among the i-th expert.

The value of T_i is calculated for each expert who has the same ranks. For example, if one expert has the highest number of repetitions, and the repetitions are distributed across ten factors as follows: 2, 1, 4, 4, 4, 4, 8, 8, 8, 10…

For example: (4p – 4, 3p – 8).

$$T_i = \frac{1}{12}(4^3 - 4 + 3^3 - 3) = 7.$$

The value of W is random, so it is necessary to check its significance (i.e., whether it is truly random) using a specific criterion.

If the number of factors n is greater than 7, the χ^2 criterion can be used to check the significance of W. It has been observed that the value of m(n – 1) follows a chi-squared (χ^2) distribution with degrees of freedom g = n – 1. Therefore, it can be written as:

$$\chi^2 = \frac{m(n-1)\sum_{j=1}^{n} d_j^2}{\frac{1}{12}m^2(n^3 - n) - m\sum_{i=1}^{m} T_i} = \frac{\sum_{j=1}^{n} d_j^2}{\frac{1}{12}mn(n+1) - \frac{1}{n-1}\sum_{i=1}^{m} T_i}. \tag{3.25}$$

In order for the value of W to be significant at the selected level of significance (which in our tasks is set between 0.01 and 0.05, corresponding to a confidence probability of 0.95 to 0.99), and given the number of degrees of freedom g = n – 1, the following condition must be met: χ^2 must be greater than the tabular χ^2 (as per the table value (3.26)); furthermore, the tabular value of χ^2 is found in the χ^2-distribution table.

The input to the table consists of the number of degrees of freedom and the level of significance. If condition (3.26) is not met, then the hypothesis that the experts' agreements are not random should be rejected. When $n \leq 7$, the Fisher's criterion should be used for these same purposes.

Here's an example: 10 factors were ranked by three experts using biological specimens.

Many biological processes involve a specific interaction between DNA-binding proteins and DNA sites. The mechanism of recognition, which is specific to the structure and sequence of the molecules involved, remains poorly understood despite some successes achieved in experimental studies of mutations and in computer analyses of known X-ray structures of protein-DNA complexes. One possible reason for this lag could be a lack of data. When analyzing the experimentally determined structures of protein-DNA complexes, a number of patterns can be identified: preferentially, alanine-thymine pairs are formed (due to the interaction between methyl groups), as well as hydrogen bonds between arginine and guanine, and asparagine and adenine. It has been shown that the contact region (protein-DNA) is enriched with polar amino acids. Purines are more selective in their binding to amino acids than pyrimidines. Aromatic amino acids may have various preferences [64].

The task assigned to the three experts is to rank the ten amino acids in order to subsequently determine the contact area between proteins and DNA. Let's denote these amino acids as: $x_1, x_2, \ldots, x_{10}$..

The results of the ranking will be summarized in a table (see Table 3.3).

Table 3.3

E|f	X_1	X_2	X_3	X_4	X_5	X_6	X_7	X_8	X_9	X_{10}
1	7.5	3	4.5	10	6	9	7.5	1	2	4.5
2	8	4.5	4.5	6.5	9	6.5	10	2.5	2.5	1
3	4.5	4.5	4.5	10	8	8	8	2	4.5	1
$\sum_{j=1}^{3} x_{ji}$	20	12	13.5	26.5	23	23.5	25.5	5.5	9	6.5
d_j	+3.5	–4.5	–3	10	6.5	7	9	–11	–7.5	–10
d^2	12.25	20.25	9	100	42.25	49	81	121	56.25	100

Suppose the ranking values assigned by the three experts for the factors were as follows:

$$\bar{X} = \frac{1}{2}m(n + 1) = \frac{33}{2} = 16\frac{1}{2} \quad (3.27)$$

$$d_j = \sum_{j=1}^{3} x_{ji} - \bar{x} \quad (3.28)$$

The agreement coefficient W is determined using formula (3.24).

$$\sum_{j=1}^{n} d_j^2 = 591.$$

$$\frac{1}{12}m^2(n^3 - n) = \frac{1}{12}3^2(10^3 - 10) = \frac{9 \cdot 990}{12} = 742.5;$$

$$T_1 = \frac{1}{12}[(2^3 - 2) + (2^3 - 2)] = \frac{1}{12} \cdot 2(2^3 - 2) = 1;$$

$$T_2 = \frac{1}{12}[(2^3 - 2) + (2^3 - 2) + (2^3 - 2)] = \frac{1}{12} \cdot 3(2^3 - 2) = 1.5;$$

$$T_3 = \frac{1}{12}[(4^3 - 4) + (3^3 - 3)] = 7;$$

$$W = \frac{591}{742.5 - 3(9.5)} = 0.828.$$

$n = 10 \geq 7$, so we use the χ^2 criterion.

The number of degrees of freedom is equal to $n - 1$, which is 9.
We determine χ^2 using formula (3.25):

$$\chi^2 = \frac{591}{\frac{1}{12} \cdot 3 \cdot 10 \cdot 11 - \frac{1}{9} \cdot 9.5} = 22.3$$

We choose a significance level of 0.2 (confidence probability of 0.8).

We find the χ^2 tabular value. χ^2 tabular = 12.2.

Since $\chi^2 > \chi^2$ tabular, the hypothesis that the experts agree in their opinions can be accepted. For the task at hand, the factors x_4, x_6, and x_7 can be disregarded when making predictions. The amino acids will then be ranked in the following order based on their sequence: x_8, x_{10}, x_9, x_2, x_3, x_1, x_5.

Therefore, if we use experts' forecasts, the first thing we need to do is to check how consistent their predictions are.

Recently, due to the increasing scope of application for machine learning methods, "expert evaluation" techniques have significantly lost their prominence. This is because new possibilities have been uncovered in the field of prediction, particularly those related to the principle of model self-organization [65].

3.11 ARTIFICIAL NEURAL NETWORKS

Neural networks have become widely used wherever prediction, classification, or control tasks need to be solved.

The greatest benefits of using neural technologies are achieved in such areas as pattern recognition, data analysis, modeling, and control [66].

Each neural network is based on relatively simple, and in most cases homogeneous, elements (cells) that simulate the functioning of neurons in the brain. Each neuron is characterized by its current state, similar to nerve cells in the brain, which can be either excited or inhibited. Theoretically, the number of layers as well as the number of neurons in each layer can be arbitrary; however, in practice, these are limited by the computer's resources or the specialized microchips on which the neural network is implemented. The structure of a neural network is chosen according to the specific requirements and complexity of the task at hand. The more complex the neural network, the more extensive the problem-solving capabilities it possesses. Artificial neural networks can adjust their responses based on external conditions. There is no doubt that neural networks achieve better results than traditional computing methods—in certain areas, at least. Neural networks are particularly well-suited for control-related tasks, as their parallel structure enables extremely fast operation. They can also be used for applications such as data analysis-based forecasting and making recommendations.

CHAPTER 4. PHILOSOPHICAL ASPECTS OF THE KNOWLEDGE OF NATURE

In the overall education of any person, their philosophical training holds particular importance. Regardless of the path they take in life, when encountering various facts, events, and experiences, people interpret them in their own way, attempt to understand their causes, and determine their own attitudes towards them. This is especially true for those who are on a path of knowledge and understanding. It is precisely here that their philosophical training becomes particularly valuable.

No matter what generalization is made in science, it always expresses some particular point of view and, ultimately, some philosophical system. Science cannot exist without philosophy. For instance, the laws of mechanics discovered by Galileo, as well as the laws governing the motion of planets around the Sun, held great philosophical significance. Galileo argued that philosophy should be studied through the "great book of the world" written by nature itself, rather than the writings of the Greek philosopher Aristotle (384–322 BCE). He believed that philosophy is recorded in this "book of the world," constantly accessible to our eyes, and that it is written in the language of mathematics.

The recognized classic of natural sciences, V.I. Vernadsky, states in [67]: "One must study philosophy, but one cannot become a philosopher merely through the study of philosophy alone; for the essential characteristic of philosophy is the inner, sincere effort of contemplation… At the heart of philosophy lies the primacy of human reason."

Thinkers from the past have shown us paths that allow us to get closer to the truth. However, their sometimes contradictory approaches do not fully resolve the issues related to our understanding of the world.

The work of a scientist consists of three parts: formulating principles, deriving logical conclusions from these principles in order to understand the relevant observed facts, and then experimentally verifying those observed facts. Thus, science begins with common sense—the intersection of the sensible and the rational. Linking the principles of science with common sense is an important task for philosophy [39].

Humans can create a model of the natural world based on direct experience, reason, and intuition. Sensations and emotions that pass through the filter of reason lead to logically sound conclusions about the properties of reality (both objective and subjective). However, some of the properties that are confirmed by practice do not fit within a model constructed solely on the basis of reason. In those areas where reason is powerless, faith comes into play—a belief that is based on nothing but sheer absence of doubt.

For a manager, sensations and feelings may serve as the starting point when forming a worldview, which is why philosophy is called subjective.

The idea of subjectivity is nothing new; philosophers have discussed it for centuries. Descartes pointed out that, at least, subjective thought itself certainly exists, expressing this in the famous statement: "I think, therefore I am" [46].

Scientific knowledge develops through a dialectical unity of processes of differentiation and integration; in each historical period, integration represents the concluding stage of this development.

Mathematical language enables us to effectively describe various phenomena in natural and social activities, and to identify different patterns of stability and invariance within these phenomena. By doing so, we can uncover the deepest laws governing nature and society. Mathematical formalism provides a solid foundation for the synthesis of various sciences.

The German philosopher Immanuel Kant sought to understand the workings of cognitive processes. As a result, he wrote the famous works "Critique of Pure Reason" [68] and "Prolegomena to Any Future Metaphysics That Might Serve as Science" [69]. He did not question where cognitive abilities came from; he simply observed that they existed and then studied how they functioned. Through this analysis, Kant concluded that there exists a system of categories, concepts, logical rules, and methods of inference that are used in the understanding of nature. This system of "pure reason" is of an a priori nature—it exists within our consciousness, prior to all experience—and it forms the basis for scientific knowledge of nature.

In modern science, concepts such as structure, symmetry, and proportion play a fundamental role; these notions originated in mathematics. The "golden ratio," from the perspective of philosophical science, represents a new relationship between opposing qualities or characteristics of an object—in other words, a quantitative ratio between two opposites. Opposites refer to the two aspects of the same entity or phenomenon that are in conflict with each other due to their absolute polarity. Philosophy has clearly and unambiguously established that the driving forces behind the development and progression of the existing world are the unity and the struggle between opposites.

Mathematics consists of a collection of abstract structures. Each such structure, up to isomorphism, is capable of representing the various forms that the material world possesses in such abundance. To understand the structure of matter is, to a large extent, to identify its corresponding mathematical structure [70].

Mathematical modeling is an essential component of the process of mathematizing knowledge, which, as we know, involves applying mathematics to non-mathematical sciences. The particular nature of philosophical inquiry requires a special type of mathematization. What are the characteristics of this type of mathematization? In search of answers to these questions, one turns to the works of V.V. Nalimov, who discusses the semantic continuum and the cosmically universal, multi-level nature of consciousness. By studying the "level of contemplation," Nalimov develops a probability-based concept of consciousness that is grounded in Bayes' theorem. He constructs and justifies a probabilistic model of meaning generation as the most suitable approach for interpreting the cosmically universal phenomenon of consciousness, demonstrating how mathematical models can be utilized in the development of philosophical thought. [71]

The mathematical model must be adequate, meaning it accurately reflects the underlying mechanisms of the system component or environment being described.

By its very nature, chance does not exist; since nature possesses complete knowledge about the current state of things, it is not dealing with a model but with the phenomenon itself. Therefore, all current manifestations of reality are, in this sense, lawful for nature.

Stanislaw Zenin, a Doctor of Biological Sciences and candidate in Chemical and Philosophical Sciences, noted: "The nature of life, just like the nature of the mind, objectively exists in our world as some form of active material. If we were able to discover the law or the universal 'program' that governs the development of our material world, then we could use it to explain the emergence of life on our planet. For now, we know that this 'life program' exists and spreads throughout all the informational-phase states of that system; depending on the conditions of a particular environment or planet, this program is realized, ultimately leading to the emergence of life." [72]

The model's flexibility with regard to information processing was a result of the incomplete coverage of the relevant issues. The interpretation of the concept of "information" can be illustrated through the following chain of definitions:

1) "Information" – refers to knowledge or data about the surrounding world.
2) "Data" (information) refer to knowledge or understanding of something.
3) "Knowledge" refers to the understanding of reality through consciousness.
4) "Consciousness" is the ability of a human being to think logically. [73]

The philosophical aspects of Bayesian theory can be roughly divided into two groups that address:

- The ratio between deductive and inductive methods of reasoning;
- This relates to the interpretation of the concept of probability.

According to H. Jeffreys' theory [74], deductive logic allows for three types of statements:

- The statement can be completely proven.
- The statement may be completely rejected.
- The statement cannot be proven nor refuted.

The last statement corresponds to Gödel's Incompleteness Theorem. The essence of the theorem is that in any sufficiently complex theory, there exist statements that cannot be proven or disproven solely using the axioms of that theory. Alternatively, it is also possible to prove any given statement, regardless of whether it is true or false [75].

To organize the process of induction, Jeffreys proposes eight rules: five main (mandatory) rules and three additional (recommended) rules [56].

There are two extremes of probability: objective probability (mainly referred to in terms of frequency interpretation) and subjective probability, which is considered a "reasonable level of confidence." In an objective sense, the probability of event A is determined by performing the relevant set of conditions [76] an unlimited number of times, assuming that the process is fundamentally reproducible.

In a subjective sense, probability is a quantitative assessment of the likelihood of an event occurring, which a researcher makes based on, for example, their own personal experience—or in other words, on prior information regarding that experience or event. Subjective probability is consistent with the theory of inductive reasoning and also satisfies the axioms of probability.

In most cases, the parameter that reflects some objective property of the object being studied is not random. However, since the value of the parameter is unknown, it is considered "uncertain." Nevertheless, using objective probabilities to describe uncertainty is not appropriate because there is no possibility of repeated experimentation under unchanged conditions. On the other hand, subjective probability, as a measure of confidence, is suitable for describing uncertainty. This ultimately leads to the rejection of objective probabilities in favor of subjective ones.

The idea of using subjective probabilities in place of objective ones, according to De Finetti, "shocked many people who viewed objectivity in the strict sense as a necessary attribute of probability and science. Nothing was lost, except for an apparent illusion. This illusion of objectivity has now been replaced by an objectively achievable level of objectivity—i.e., the degree of objectivity attained by human science through the human mind" [76, p. 121]. Cornfield [77] agrees with this view. He believes that the objectivity of science is reflected in a mathematical principle: different researchers, starting from various prior probabilities, will always arrive at the same posterior conclusions when conducting a large number of experiments. This principle has been mathematically established through theorems independently proven by Bernstein [78] and Mises [79]. These statements are in line with the materialist dialectic, which holds that the sole criterion for truth (and therefore objectivity) is practice, understood in a statistical sense as a collection of numerous experimental data. Otherwise, discussing objectivity and claiming to derive "ultimate truths" would be meaningless.

The existing disagreements between "objectivists" and "subjectivists" seem to be mainly terminological in nature. The issue is that, when discussing probability, people usually refer to the frequency [51].

The Polish science fiction writer Stanisław Lem, in his book "Summa Technologiae" (1964), which discusses issues related to cybernetics, astronautics, and futurology, stated in Chapter 7: "Science has entered a competition with Nature. Although it wins one battle after another, it allows itself to be so deeply involved in the consequences of those victories, and it exploits each of them so thoroughly that instead of strategy it resorts to tactics. The paradox is that the more successes there will be in the future, the more difficult the situation will become, because—as we have already shown—not everything we acquire can always be effectively utilized." This is his interpretation of the past and the future.

Living nature is a part of the universe, and since all processes within it are interconnected, the primary task of the mind is to understand the laws that govern the world. Knowledge represents the eternal pursuit of the human mind [80].

The main principle for evaluating the credibility of scientific results is the principle of correspondence [7]. This principle states that any credible scientific finding must not contradict scientific knowledge that meets generally accepted criteria of truth, as well as empirical facts.

The first part of this principle has been the subject of intense criticism for many decades by highly authoritative scientists. For example, L. Wittgenstein argued that if new scientific knowledge does not contradict any elements of established theories, then it is merely tautological—meaning it is a rephrasing of existing results in different terms [81]. Therefore, according to L. Wittgenstein, such knowledge does not meet the criteria for being considered truly novel.

It is necessary to emphasize the fundamental importance of L. Wittgenstein's position. This position may not always be convenient from a practical standpoint, but in essence it is correct. It represents the dividing line between two fundamentally different philosophical approaches to the issue of understanding the scientific world. From the perspective of dialectical philosophers (successors to Hegel and their interpretations of his teachings), the primary source of knowledge about objects is the subject [81, 82].

The subject's knowledge is stored in its memory, books, magnetic recordings, and so on. From this, we derive the thesis: "Read the classics! They have everything." In the 1930s, L. Wittgenstein demonstrated that by developing the knowledge of other subjects, one can only arrive at a tautology. To make fundamental progress, it is necessary to reject and prove the falsity of certain pieces of previously acquired data and knowledge that do meet generally accepted criteria of truth. This position is held by the philosophical schools of critical realism, constructivism, and intuitionism [81, 83–88]. Their main thesis is: "The primary (but not the only!) source of knowledge about an object is the object itself."

A philosophical interpretation of the methodological problems in genetics, from the perspective of materialist dialectics, allows for a clearer understanding of the objective dependence of scientific methodology in the analysis of the processes of formation and development of concepts and theories in modern genetics [89].

CHAPTER 5. THE STRUCTURE OF DNA AND THE GENETIC CODE

5.1 A BRIEF HISTORICAL OVERVIEW OF THE DEVELOPMENT OF GENETICS KEY CONCEPTS AND DEFINITIONS

Genetics is a rather complex and mysterious science. The genetic code in particular holds many secrets. To make more thorough use of the mathematical apparatus, let us take a brief excursion into this science.

Genetics is the science that studies the laws of heredity and variation in organisms, as well as the methods for manipulating these phenomena. The philosopher I.T. Frolov noted: "The history of genetics, from the research of Gregor Mendel to its contemporary developments, contains many mysterious and astonishing aspects that require a philosophical approach; such aspects not only stimulate philosophical inquiry but also inspire imagination and arouse a keen interest in scientific investigation." [23]

The discovery made by Gregor Mendel, which was ahead of its time and was published in 1865, attracted little interest from his contemporaries. It was only many years later, in 1900, that researchers such as Hugo De Vries (Netherlands), Carl Correns (Germany), and Erich Chermak (Austria) rediscovered Mendel's laws and demonstrated their validity. Thus, Mendel's laws laid the foundation for scientific investigations conducted by scholars from various countries.

The founder of the theory of heredity was the Austrian natural scientist Gregor Mendel, who, based on experiments with crossbreeding pea varieties (1856–1863), formulated the laws governing the independent segregation of traits and the combination of genetic factors.

The contribution of G. Mendel lies in the fact that he found a simple way to express the types of inherited traits observed in pea crossbreeding, as well as their numerical relationships, in mathematical formulas. He traced the inheritance of these traits over several generations [46].

Mendel took two varieties of sweet peas that differed in one or more traits (for example, the seeds of one variety were green, while those of the other variety were yellow), and crossed them. In the first generation, all the seeds of the offspring plants were yellow. This color dominated the green trait; therefore, Mendel referred to this trait, which was expressed in the hybrids of the first generation, as "dominant." The green trait was accordingly called "recessive." One might think that since all the plants in the first generation had yellow seeds, the second generation would also have only yellow seeds. However, this did not happen. In the second generation, crossbreeding between different plants no longer took place (Mendel used self-pollinating plants); nevertheless, plants with both yellow and green seeds appeared. On average, out of every four plants, three had yellow seeds and one had green seeds. By conducting numerous experiments and performing thorough quantitative analyses, Mendel determined that a ratio of 3:1 was characteristic of the inheritance of these traits.

The inheritance of traits in the 3:1 ratio was termed "phenotypic segregation." In genetics, a phenotype refers to the complete set of characteristics of an organism.

Mendel translated the biological law into mathematical language [90].

The unit of heredity, the gene, has taken physical form in a portion of the deoxyribonucleic acid (DNA) molecule; this DNA portion determines the development of a particular trait.

The understanding of how genes and traits interact is summarized by the gene-to-trait framework, in which genes control the development of traits.

In 1868, the Swiss researcher F. Miescher discovered a new class of biologically important organic compounds that contained carbon, nitrogen, and phosphorus. He named this substance "nuclein" (from the Latin "nucleus," meaning "core" or "center"). The compound isolated by F. Miescher was later named nucleic acid in 1899. Therefore, F. Miescher is the discoverer of DNA [46].

DNA is the hereditary material of a cell, but there are other forms of transmitting genetic information as well. Information can also be carried by cellular structures that are themselves inherited.

The main components of a cell are: the membrane, cytoplasm, and nucleus [46].

Most of the DNA is found in the cell nucleus. The cell was discovered by the English natural scientist Robert Hooke in 1665. Living organisms that have a cellular structure can be divided into two main types: Prokaryotes (organisms without a nucleus), to which bacteria belong; Prokaryotic cells do not possess a nucleus or membrane-bound organelles. The second type of organism is Eukaryotes (organisms with a nucleus); this group includes both single-celled and multi-celled organisms.

In a prokaryotic cell, there are 2,000 to 3,000 different proteins; in eukaryotes, this number exceeds 50,000, of which more than 2,500 possess catalytic activity—i.e., they are enzymes.

Each cell generates an electric current. In experiments, muscle and nerve cells have long been used, as they most clearly demonstrate the effects of electrical currents. A nerve cell is a tiny "battery" that is capable of generating an electrical impulse, which travels along its extensions without fading away; this impulse can cover distances of hundreds of centimeters. [91]

Moreover, many animal organs generate strong electrical fields that can extend beyond the boundaries of the organ itself. The most well-known example of this is the heart; recordings of the electrical currents in the heart (electrocardiography) have become a routine medical practice. Electricity can also be emitted from an organism's body; this is the case with electric eels and electric rays, which generate currents with voltages exceeding 200 volts [92].

Living cells consist largely of water. Water is a weakly magnetic material: it becomes magnetized in the direction opposite to the applied magnetic field, that is, it is a diamagnet [93].

Upon microscopic examination of the cell, intensely staining bodies can be observed within the nucleus. These bodies appear only during the cell division process, disappear afterwards, and reappear before the next division. E.A. Kaminskaya [94] notes: "It seems that they become invisible for some time. These structures were extensively studied by W. Waldeyer in 1888; due to their intense staining properties, they were named 'chromosomes.'" Chromosomes were therefore studied even before scientists discovered the structure of DNA, as they could be observed with a regular microscope. One of the main functions of chromosomes is transcription—the process of synthesizing molecules of informational RNA (messenger RNA) from genes. Each chromosome consists of DNA and protein that are bound together.

Proteins and nucleic acids that make up a chromosome are composed of atoms that occupy specific positions in Mendeleev's Periodic Table of Elements. DNA and RNA consist only of hydrogen, oxygen, carbon, nitrogen, and phosphorus (H, O, C, N, P). The polypeptide chains that make up proteins are also composed only of hydrogen, oxygen, carbon, nitrogen, and sulfur (H, O, C, N, S). These sets of elements are characterized by the following properties:

1. They are very short—there are only six elements in total.
2. Out of the five elements in each set, four are identical.
3. These six elements are compactly arranged within Dmitri Mendeleev's Periodic Table of Elements.
4. None of these elements are metals.
5. All of these elements have a small number of protons in their nuclei.

This means that periodicity at the atomic level [29] also played a role in defining the directed development of the structure and function.

Chromosomes do not obey the laws of gravity, magnetism, probability theory, or natural selection. An organism or a cell may die, but a chromosome will survive due to its adaptability.

Every phenomenon that, when studied at the level of cells and chromosomes, initially appeared to be the result of a random process turned out to be highly organized when examined at the molecular level.

The transmission of inherited traits, the occurrence of mutations, the exchange of chromosome segments, the rearrangements within and between chromosomes, as well as many other genetic phenomena, all have a molecular basis. At the chromosomal level, this molecular organization exhibits a clear degree of order. Within chromosomes and cells, this order is primarily regulated by a set of RNAs and proteins that arrive at the right place at the right time [29].

The information regarding the organization of chromosomes indicates that even their simplest structural components possess an orderly nature. The functional repeating unit of chromosomal material is the nucleosome. The nucleosome is the primary structural unit of a chromosome. It consists of a histone core around which DNA is wound, forming a length of 145 base pairs (bp) and making 1.75 turns [95]. In the cells of higher organisms, there are between 50,000 and 500,000,000 nucleosomes, which are clearly visible under an electron microscope. The arrangement of nucleosomes along a chromosome is of great importance, as they prevent the interaction between DNA and proteins (including DNA-binding proteins). Nucleosomes are distributed periodically throughout the chromosome in a so regular manner that their distribution can be predicted. This means that nucleosomes are not formed on DNA at random sites [96].

In 1953, J.D. Watson and F. Crick constructed a model of DNA and explained the mechanism by which information is encoded through the sequence of deoxyribose molecules, phosphoric acid, and nitrogenous bases. The model of DNA structure was presented as a double helix that is tightly coiled around a protein backbone. Each base pair is twisted approximately 36 degrees around the axis of the helix with respect to the next base pair; therefore, each turn of the helix contains 10 base pairs of the form [97].

A gene is a portion of the DNA sequence that contains information about the amino acid sequence of a particular protein.

Nucleotides are the fundamental units of genetic material; they represent the monomeric components of the polymer molecule DNA.

The credit for establishing the pattern of genes on chromosomes goes to the American biologist T. Morgan (1866–1945), who experimentally supported the chromosomal theory of heredity (Nobel Prize in 1933).

A gene contains information that controls the development of a particular feature.

In 1956, it was finally established that the human karyotype consists of 23 pairs of chromosomes (a diploid set): 22 pairs of autosomes and one pair of sex chromosomes (XX for females, XY for males). In other words, the combination of two X chromosomes results in a female individual, while the combination of one X chromosome and one Y chromosome leads to the formation of a male individual.

The following levels of organization were identified: gene – chromosome – nucleus – cell – feature.

It was generally accepted that DNA itself does not participate in the synthesis of proteins; rather, it merely regulates protein production via RNA. In 1958, F. Crick formulated a fundamental principle of molecular biology, which was later termed the "central dogma of molecular biology." According to this principle, the pathway for information transfer is as follows:

DNA ------→ RNA ------→protein.

5.2. THE STRUCTURE OF DNA AND RNA

When a DNA molecule is broken down into its constituent parts, the following are obtained: nitrogenous bases (adenine, guanine, cytosine, thymine), a sugar (referred to as ribose), and a residue of phosphoric acid. DNA is a polymer, which means it has a chain-like structure. In addition to DNA, there is another substance called ribonucleic acid (RNA) that has a similar chemical composition to DNA. RNA serves as a link between DNA and proteins. RNA is a high-molecular-weight organic compound that belongs to the nucleic acids; its nucleotides include adenine, guanine, cytosine, uracil, and ribose. The difference between DNA and RNA lies in the fact that thymine replaces uracil in DNA, and deoxyribose replaces ribose in RNA. Ribose contains one more atom of oxygen, which is why the sugar portion is called "deoxyribose" (i.e., "ribose without oxygen"). DNA exists in the form of two complementary strands that are twisted around each other in a spiral structure. DNA has a crystalline structure that functions as a receiver, transmitter, and amplifier of frequencies (or "light" signals). In terms of its shape—the giant double-helical structure—DNA is an ideal electromagnetic antenna: it is elongated, resembling a fan-like structure that efficiently captures electrical vibrations; moreover, when viewed from above, it has a ring-like shape, making it an excellent magnetic antenna. Since humans have inherited genetic information since ancient times and all members of the same species share identical DNA sequences (the [39] wavelength), they are capable of transmitting and receiving information using this same frequency range.

Within the double helix structure, adenine forms two hydrogen bonds with thymine, while guanine forms three hydrogen bonds with cytosine. In DNA, these bonding patterns determine the specificity of the base pairs (A = T and G ≡ C). The two dashes in the diagrams represent the two hydrogen bonds in the A=T pair, and the three dashes represent the three hydrogen bonds in the G≡C pair. The strands of the DNA double helix are oriented in opposite directions, meaning they are antiparallel to each other. The structure and biological properties of nucleic acids are determined by their chemical characteristics, the quantitative ratio of the nitrogenous bases, as well as the sequence of the corresponding nucleotides.

Nucleotides are the constituent units of nucleic acids. They consist of a nitrogenous base (purine or pyrimidine), a sugar (ribose or deoxyribose), and a phosphate group. Biochemists represent all five nitrogenous bases in scientific books and journals using uppercase letters from the Russian or Latin alphabets: Adenine is denoted as A, Guanine as G (in the Latin alphabet), Cytosine as C, Thymine as T, and Uracil as U. The structure of the double helix of a DNA molecule is illustrated in Figure 5.1.

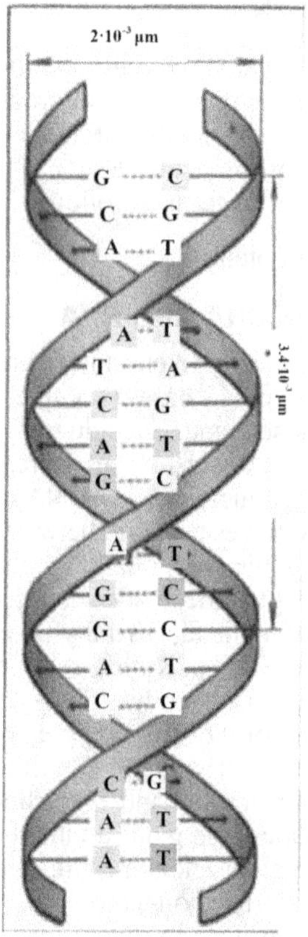

Figure 5.1. Diagram of the DNA double helix molecule

In DNA, sugar alternates with phosphoric acid, and one nitrogenous base is attached to each sugar. The alternation of sugar and phosphate is ordered, whereas the bases appear to alternate in complete disorder [98]. This triumvirate of sugar, phosphate, and base was named a nucleotide. In all, DNA contains four nucleotides — one for each of the different nitrogenous bases.

Of these four bases, two are purines: adenine and guanine, and two are pyrimidines: cytosine and thymine.

The Watson-Crick model of DNA structure took into account a number of previous discoveries. One of these was the rules established by E. Chargaff in 1951 regarding the proportion of purine and pyrimidine bases in DNA molecules of different species. According to these rules, the amount of adenine in a DNA molecule is always equal to the amount of thymine (A = T), and the amount of guanine is equal to the amount of cytosine (G = C). Additionally, as mentioned earlier, the total number of purine bases equals the total number of pyrimidine bases (A + G = T + C). Any DNA molecule can be characterized by the ratio (G + C) / (A + T), which is constant for a given species and varies between 26–74 percent by mole.

DNA has a regular helical structure with a diameter of 20 angstroms (1 angstrom = 10^{-8} cm); each full turn of the helix occurs every 34 angstroms (3.4 nm), meaning that there are 10 nucleotides in one turn. The scale of measurement is as follows: 1 nm = 10^{-9} m. The double helix is right-handed; when viewed along the helical axis, the turns proceed in a clockwise direction. The size of DNA molecules is expressed in various units: picograms (pg), daltons (1 D = mass of a hydrogen atom = 1.67×10^{-24} g), or nucleotide pairs (ntp). Specifically, 1000 nucleotide pairs constitute 1 kilobase (kb). The conversions between these units are as follows: 1 pg = 10^{-12} g = 0.6×10^{12} D = 0.9×10^{9} ntp. The terminal region of DNA, called the telomere, is a region 3–20 kb in size, consisting of a sequence of TTAGGG repeated in tandem; various specific proteins bind to this region, and this complex is referred to as the telomere [99]. Measurements of the genome, i.e., the amount of DNA in the haploid set of chromosomes, have revealed the following: The genome of Escherichia coli, a favorite organism for molecular genetics studies, consists of a single circular DNA molecule with a nucleotide count of 4.6×10^{6} ntp and a length of 1.4 mm. The human haploid genome (often referred to as the "crown of nature" or the "measure of all things") contains 3.2×10^{9} ntp of DNA, which is equivalent to 3.5 pg of DNA. In a diploid cell nucleus, this amount represents 7 pg of DNA. This constitutes less than 1% of the weight of a human cell, whose weight averages around 1000 pg.

The analysis revealed [99] that in a human body weighing 70 kg, there is 45.5 kg of oxygen (65%), 12.6 kg of carbon (18%), 7 kg of hydrogen (10%), 2.1 kg of nitrogen (3%), 1.4 kg of calcium (2%), 0.7 kg of phosphorus (1%), and approximately 0.7 kg of all other elements combined (mainly potassium, sulfur, sodium, chlorine, magnesium, iron, and zinc) (1%).

The differences in the composition of RNA compared to DNA have important biological implications. RNA molecules can only perform their functions in their single-stranded state. This is particularly evident during the process of translating mRNA on ribosomes [95]. The bases in one strand are connected by hydrogen bonds to the bases in the other strand. All bases in both strands are paired with each other: adenine can only pair with thymine, and guanine can only pair with cytosine.

The main characteristics inherent in living phenomena were identified by the prominent theorist B.M. Mednikov (1982), who summarized the axioms of theoretical biology. One of these axioms, namely the second axiom proposed by

N.K. Koltsov (1872–1940), states: "The genetic program is established according to a matrix principle; the gene of the preceding generation serves as a template for the formation of the gene of the subsequent generation." [95].

The DNA molecule cannot serve as a template for protein synthesis. This conclusion follows logically from the fact that nearly all of the DNA is found in chromosomes, which are located within the nucleus, whereas protein synthesis takes place in the cytoplasm of the cell. DNA programs the synthesis of RNA. First, through transcription of genetic information from DNA, a special informational (messenger) mRNA is formed. Synthesis proceeds with the aid of the enzyme RNA polymerase. A copy in the form of an RNA molecule is taken from a DNA region using the enzyme RNA polymerase.

A paired molecule of DNA looks like this:

```
           ←——————————————————————————
3′ .....   G   A   G   C   T   G   C   T  ..... 5′
           |||  ||  |||  |||  ||  |||  |||  ||
5′ ....    C   T   C   G   A   C   G   A  ....3′
           ——————————————————————————→
```

These base pairs are called complementary bases, and the strands of DNA are referred to as complementary strands. The long purine bases (A and G) pair with the short pyrimidine bases (C and T). In general, these base pairs can be quite easily separated. When the hydrogen bonds between them are broken, the DNA strands separate into two single-stranded molecules. Each of these single-stranded molecules is then extended by adding free nucleotides that are present in the cell. The rules for how these nucleotides are paired remain the same: A pairs with T, and G pairs with C. The newly added nucleotides not only bind to their corresponding partner bases but also to each other, forming a new DNA chain (see Figure 5.2).

When the strands of the original pair separate completely, there will be two pairs of nucleotides in the cell instead of just one. The sequences of the base pairs that make up these molecules will be identical to each other. This process is called replication. During self-replication, a double-stranded DNA molecule is split into two strands, and each of these strands serves as a template for the synthesis of its complementary counterpart.

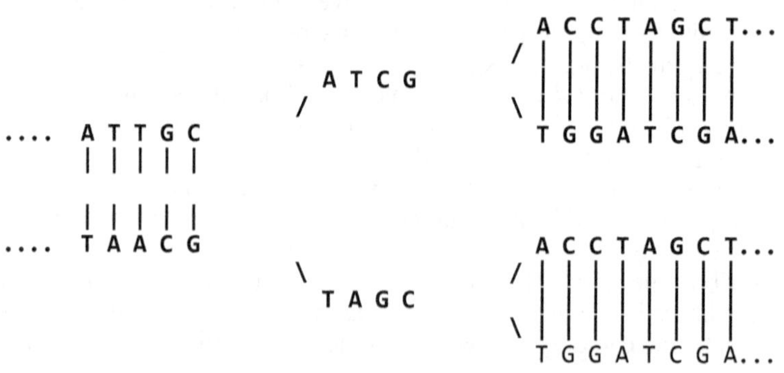

Figure 5.2. The separation of DNA strands during the replication process

DNA consists of two strands: one strand contains information about the structure of a protein (the "coding" strand), while the other strand has no meaningful content. The coding strand serves as a template for the synthesis of RNA, with the assistance of an RNA polymerase enzyme. It is unlikely that both DNA strands would encode the same protein structure. The sequences of nitrogenous bases in the two strands are oriented in opposite directions; therefore, the amino acid composition of the proteins synthesized from these strands will be different (see Figure 5.3).

DNA → Transcription → mRNA → Translation → Amino Acids

From left to right

CCGACA --------------→ GGCUGU ------------→ Gly —Cys

GGCTGT --------------→ ACAGCC ------------→ Thr —Ala

From right to left

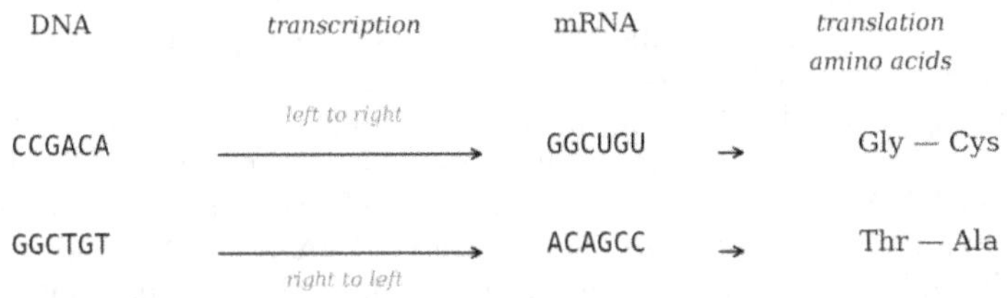

Figure 5.3. Directions of the nitrogenous base sequences

When DNA is first copied, the resulting RNA contains both exons (which can encode amino acids) and introns (which do not encode amino acids). Subsequently, cellular mechanisms remove the introns entirely or partially (this process is called splicing). The resulting messenger RNA (mRNA) molecules can then convey information to various proteins derived from the same gene. This mechanism can be illustrated using the diagram shown in Figure 5.4 [100].

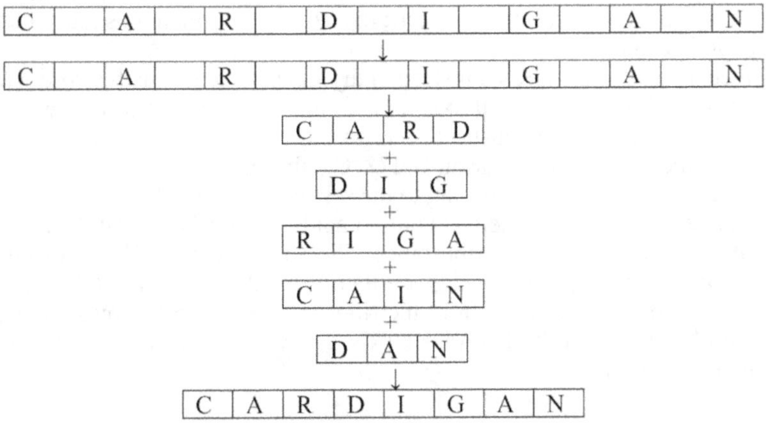

Figure 5.4. Schematic illustration of the splicing process

At the top of this diagram, a DNA molecule is shown. The exons, which encode the amino acid sequences, are located in the cells labeled with letters. The introns, which do not participate in encoding the amino acid sequences, are represented by empty cells. For ease of understanding, all exons and introns are depicted as having the same size in the diagram; however, in reality, their sizes can vary considerably.

When DNA is first copied into RNA (as indicated by the first arrow), the RNA contains both exons and introns. Subsequently, cellular mechanisms remove the introns entirely or partially. The resulting messenger RNA (mRNA) molecules can convey information to various proteins derived from the same gene; this is illustrated in the diagram using different words. Sometimes, parts of two genes that encode different proteins combine to form a code for a new protein. This new protein will be different from the original proteins.

During splicing, some exons may also be removed. These exons will be retained in the final mRNA, while others will be lost. The various proteins produced as a result may have quite similar functions or may differ significantly from each other. A cell can express different proteins depending on its current needs or in response to the different signals it receives.

According to the RNA "text" in the ribosomes, a protein (an enzyme) is synthesized. Ribosomes are aggregates composed of protein molecules and ribosomal (rRNA) molecules. The structure of rRNA (which is also copied from DNA) is similar to the structure of messenger RNA (mRNA). The ribosome passes the mRNA strand through itself, reading the encoded instructions transmitted from DNA in the form of nucleotides, and synthesizes the corresponding amino acid sequence to form the protein. The ribosome moves along the mRNA, translating the genetic code into a different "alphabet" consisting of twenty different amino acids; in the protein "text," these amino acids serve as the "letters." The twenty amino acids are referred to as the standard set. The group of nucleotides that corresponds to a particular amino acid is called a codon. Codons are classified as "sense" if they encode an amino acid, and as "nonsense" if they do not code for any amino acid. Once the instructional text is obtained, it must be decoded. The role of the decoder in the cell is performed by a third type of RNA molecule: transfer RNA (tRNA). These RNA molecules serve as templates for protein synthesis; the genetic information is translated into polypeptide chains [2].

Since uracil is used in RNA instead of thymine, the letters in the words (which represent RNA molecules) will be A, G, C, and U. The process of transcribing instructions into RNA language occurs in the same way as DNA replication, following the rules of A-U pairing and G≡C pairing.

The rewriting is done according to the scheme shown in Figure 5.5.

Transfer RNA (tRNA) delivers amino acids to the ribosomes; the assembly of proteins takes place on the ribosomes. There is a specific tRNA molecule for each amino acid. The amino acids are then separated from the tRNA, and the completed protein is released from the ribosome. Figure 5.6 shows the adapter RNA molecules (transfer RNAs, or tRNAs), with each molecule having an attached amino acid.

	G	T	T	A	A	T	… a DNA chain
The RNA chain is:	C	A	A	U	U	A	

Figure 5.5. Transcribing information into RNA language

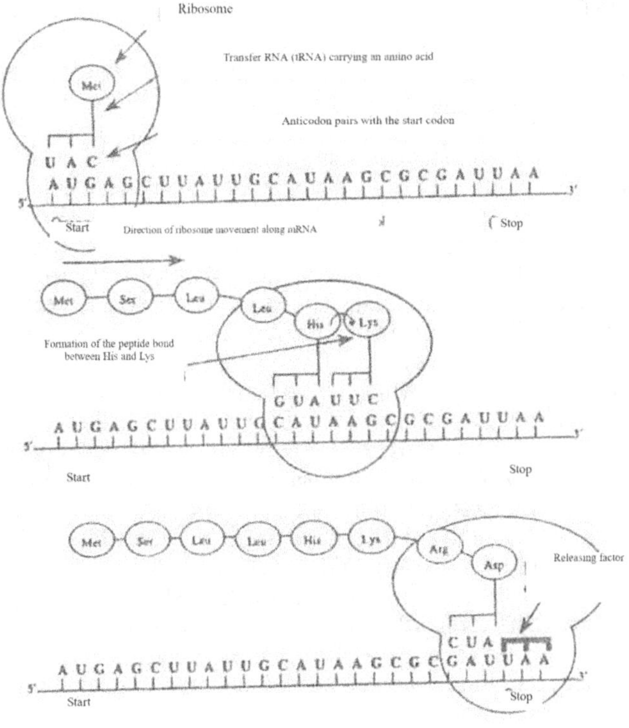

Figure 5.6. RNA adapter molecules

Within the ribosome, these amino acids bind to their corresponding codons on the mRNA via their anticodons. Thus, the sequence of amino acids encoded by the mRNA is determined by the rule that each codon corresponds to a specific amino acid as specified by the genetic code. The ribosome "reads" the linear sequence of codons, bringing together tRNA molecules that already carry their attached amino acids, so that a strong chemical bond forms between adjacent amino acids, known as a peptide bond. Once the stop codon is encountered, a termination factor binds to the ribosome, and the newly synthesized protein is released into the surrounding environment. The ribosome and mRNA can then be reused to produce the next molecule of protein [27].

5.3. THE GENETIC CODE AND ITS CHARACTERISTICS

The genetic code is a unique system specific to living organisms for "recording" hereditary information in nucleic acid molecules, in the form of sequences of nucleotides. A particular gene is responsible for the synthesis of a corresponding protein.

The idea that the information necessary for protein synthesis is encoded in genes was first proposed by the American astrophysicist G.A. Gamow, who was born in Russia and worked in Germany, Denmark, and the United States. He died in the United States in 1968.

He was the first to formulate the problem and propose a model for the genetic code in 1954. Upon learning from Crick and Watson that there were twenty amino acids, Gamow posed the question: how could the twenty amino acids be encoded using just four types of nucleotides? Initially, Gamow approached this problem purely from a combinatorial perspective, identifying issues related to the structural arrangement of the code. The answer to the question of the number of possible nucleotide combinations was straightforward: there were four types of nucleotides (A, G, T, C), and from these four, only 16 different pairs could be formed: AA, AG, AT, AC, GA, GG, GT, GC, TA, TG, TT, TC, CA, CG, CT, CC. This was insufficient to encode the twenty amino acids. Therefore, each codon must consist of at least three nucleotides. Gamow reasoned that if three nucleotides were enough, nature would likely not require more; thus, the code must be triplet-based—three nucleotides per amino acid. By adding, in turn, each of the 4 nucleotides to each of the 16 possible combinations, a total of 64 combinations were obtained, which exceeded the number of amino acids. Gamow now had to choose between two hypotheses: either the code was degenerate, or the code was determined not by an ordered triplet of nucleotides but in some other way. Guided by this same principle of efficiency, Gamow assumed that nature had assigned only one codon per amino acid, meaning the code could not be redundant. Since there were twenty amino acids, twenty different combinations of three nucleotides each had to be found. Combinatorial analysis suggested that the number 20 could be achieved by considering the states of the nucleotides without regard to their order. Gamow implemented this idea in the form of a rhombic code. However, Gamow's code was not validated by experimental evidence. Purely combinatorial attempts to decipher the genetic code were ultimately unsuccessful. [101]

The problem of the genetic code revolved around the relationship between the sequence of nucleotides in DNA and the corresponding amino acids. First, there was the question of the relative information content of DNA compared to that of proteins; second, there were the quantitative relationships between nucleotides and amino acids, which were ultimately determined by the interactions between molecules during protein synthesis. Third, there was the issue of the physical significance of the genetic code itself—the degree of randomness in this correspondence. Gamow and his colleagues gradually concluded that the coding unit was the triplet, namely a sequence of three nucleotides. Three nucleotides in a codon are more than sufficient to encode an amino acid. The specific order of these nucleotide triplets constitutes the genetic code. Genetic information is stored in DNA molecules as a sequence of four types of simple atomic groups, known as "bases": adenine, guanine, thymine, and cytosine.

Gamow was interested in how this information is translated into a sequence of 20 amino acids, which constitute protein molecules. He concluded that twenty could be obtained from four by counting the number of all possible triplets formed from four different entities. "For instance," says Gamow, "take a deck of playing cards in which we only pay attention to the suit of each card." How many triplets of the same type can be obtained? Of course, there are four suits: three of hearts, three of diamonds, three of clubs, and three of spades. How many triplets are there in which two cards have the same suit and one card has a different suit? Let's say we have four options for the third card. So we have 4 × 3 = 12 possibilities in total. In addition, we have four sets of triplets, each containing all three different cards. So, 4 + 12 + 4 = 20, and this is exactly the number of amino acids we wanted to obtain. [102] Gamow quite casually used card-game terminology in his writings: "For example, let's take a deck of playing cards…" or "Suppose we are playing a form of 'simple poker'" [Gamow, 1958]. Indeed, there are four types of bases: two black ones with legs (purines) and two red ones without legs (pyrimidines). For example, if we continue with Gamow's idea and represent adenine as ♠ (spades), guanine as ♣ (clubs), thymine as ♦ (diamonds), and cytosine as ♥ (hearts), then the sequence of amino acids would look as depicted in the figure. 5.7.

♣♠♣	♥♦♣	♣♦♠	♠♣♦	♥♣♠	♥♦♠	♦♠♦	♣♥♠	♦♠♦
Glu	Leu	Val	Ser	Arg	Leu	Tyr	Ala	Tyr

Figure 5.7

Nature seems to be playing a kind of "simplified poker" game with theorists; it's a risky gamble, and the outcome represents the greatest discovery of the 20th century. V.A. Ratner points out that the card-related incident helped bring the issue to the fore, as the theory of probability likely originated from the practice of card games and roulette [102]. Maps were among the early inventions to possess a clear, cosmological symbolic meaning. The 52 cards correspond to the number of weeks in a year; each of the four suits contains 13 cards, representing the 13 lunar months. The four suits also symbolize the elements of the universe: the red suits are associated with the warmer seasons and the forces of light, while the black suits represent the colder seasons and the forces of darkness. Cards represent a combinatorial semiotic system that signifies the various principles, plans, and elements of the universe in their entirety and within their regular interrelationships—as well as the struggles and interactions between them. So it's not surprising that G.A. Gamow used the cards in his considerations [39].

In 1954, Gamow proposed the existence of twenty specific "pockets" on the surface of the DNA macromolecule, each of which could accommodate a side chain from only one of the twenty amino acids that make up proteins. Gamow believed that each of these "pockets" was formed by a particular arrangement of four adjacent purine and pyrimidine residues located on the two strands of DNA—the two complementary polynucleotide chains. Accordingly, he suggested that DNA could serve as a direct template for the orderly assembly of specific polypeptides from these amino acid building blocks. However, Gamow soon realized that there were internal contradictions in his model; moreover, he noticed that proteins were likely not synthesized directly on DNA. Based on this realization, he revised his hypothesis. The essence of Gamow's revised model can be summarized as follows:

1. Three consecutive nucleotide pairs in DNA polypeptide chains encode one amino acid. Therefore, there are 4^3, or 64, possible triplets

(combinations of three nucleotides) that can represent the standard set of 20 amino acids.

2. Each nucleotide pair is involved in the coding of three amino acids. Therefore, the code is overlapping.
3. An amino acid can be represented by more than one type of nucleotide triplet, or word. Therefore, the code contains synonyms [103].

The period of study concerning the genetic code had begun. By 1964, most of the correlations between three-letter codons and amino acids had been established; the result of this research was what we now know as the genetic code.

F. Crick and his colleagues proposed the hypothesis of a non-overlapping code. According to this hypothesis, each triplet of nucleotide pairs in double-stranded DNA was supposed to encode one amino acid, and each base pair was only involved in determining one specific amino acid. However, this code had its drawbacks: it was unclear how the reading mechanism would know which triplet to "interpret" or decode.

G. Stent and R. Kalindar refer to the sequence [103] as an example of a nucleotide sequence: …, ATT, GCA, TCG, TGG, …, as shown in Figure 5.8.

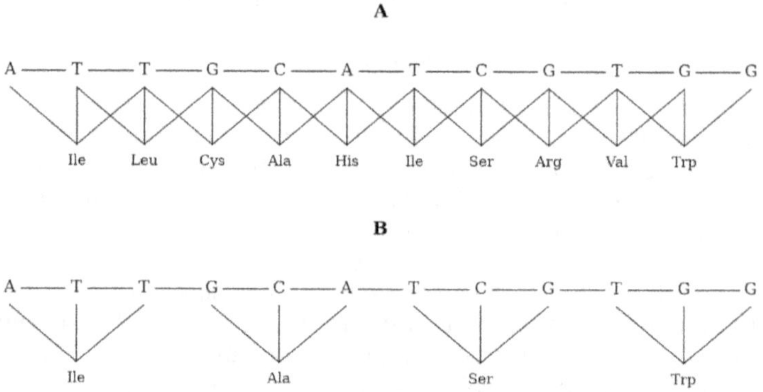

Figure 5.8. The two main types of codes that link the nucleotide sequence in DNA to the amino acid sequence in polypeptides. In both cases, one amino acid is encoded by three consecutive nucleotides

A. Overlapping code: In this type of code, each nucleotide is involved in determining the sequence of three amino acids. The code relationship, or the ratio of the number of nucleotides in a DNA segment to the number of amino acids in the polypeptide chain encoded by that segment, is 1:1.

B. A non-overlapping code in which each nucleotide is involved in determining one particular amino acid; in this case, the codic relationship is 3: 1 (Stent G.S., Molecular Biology of Bacterial Viruses, W.H. Freeman and Company, San Francisco, 1963) [103]

In Figure 5.8, "ATT" represents one amino acid, "GCA" represents another amino acid, and so on. How can this sequence be read correctly if the commas are removed? To address this issue, additional constraints were introduced to the code: each triplet such as "ATT", "GCA", "TCG", "TGG" can be automatically identified as a "word" using a dictionary of meaningful sequences. These meaningful sequences must be arranged in such a way that if any two adjacent triplets are read incorrectly, no valid amino acid code will result. Only the triplet combinations that correspond to actual amino acids in the dictionary are considered meaningful; all other possible combinations (those that may arise from incorrect readings of adjacent meaningful triplets) do not encode any amino acid and are therefore meaningless (i.e., "nonsense"). For example, if "ATT" and "GCA" are in the dictionary, then "TTG" and "TGC", which could be misread from the sequence "ATTGCA", must be considered meaningless. Thus, it is possible to create a code without commas that allows for correct information reading without the need for any starting mechanism at the beginning of the polypeptide chain or for commas between adjacent meaningful triplets. Such a comma-free code is not only theoretically feasible; in any such code, no more than 20 out of the 64 possible triplets can be meaningful, and at least 44 of the remaining triplets will necessarily be meaningless. As a result, there will be no synonyms in this type of code, and each amino acid will be represented by only one unique code word. This hypothesis nicely explains the apparent excess of possible triplets given the number of existing amino acid types. The limited nature of the dictionary ensures that the beginning and end of each meaningful sequence can be automatically identified, eliminating the need for commas or intermediate tokens [103].

If we consider the genetic code to be overlapping in nature, then when one nitrogenous base is replaced in a protein sequence, at least two amino acids must also be altered.

ATGC T→C → ACGC.

Non-overlapping genetic code implies a semantic discreteness of the code groups; as a result, the substitution of one base can lead to the alteration of only one amino acid in the protein chain.

ATG CAT (when replacing T by G) → AGG CAT.

This hypothesis is supported by experimental research, which has shown that in a protein, only one amino acid may be altered (for example, in the case of sickle cell anemia).

The final result of the code decoding process is presented in the form of a table called the "genetic code" (Figure 5.9).

The first base of each triplet is indicated in the table on the left, the second base is indicated at the top, and the third base is indicated on the right. The amino acids are represented by abbreviations corresponding to three nucleotides (A for adenine, U for uracil, G for guanine, C for cytosine). The codons AUG and GUG serve as initiation codons, marking the "start" of the protein synthesis process.

The principle behind the construction of this table—whose significance for biology is comparable to that of Dmitri Mendeleev's Periodic Table of Chemical Elements for chemistry—was proposed by Francis Crick.

	2nd position				
1st position	U	**C**	**A**	**G**	**3rd position**
U	UUU Phe UUC Phe UUA Leu UUG Leu	UCU Ser UCC Ser UCA Ser UCG Ser	UAU Tyr UAC Tyr UAA Stop UAG Stop	UGU Cys UGC Cys UGA "Stop" UGG Trp	U C A G
C	CUU Leu CUC Leu CUA Leu CUG Leu	CCU Pro CCC Pro CCA Pro CCG Pro	CAU His CAC His CAA Gln CAG Gln	CGU Arg CGC Arg CGA Arg CGG Arg	U C A G
A	AUU Ile AUC Ile AUA Ile AUG Met (Start)	ACU Thr ACC Thr ACA Thr ACG Thr	AAU Asn AAC Asn AAA Lys AAG Lys	AGU Ser AGC Ser AGA Arg AGG Arg	U C A G
G	GUU Val GUC Val GUA Val GUG Val (Start)	GCU Ala GCC Ala GCA Ala GCG Ala	GAU Asp GAC Asp GAA Glu GAG Glu	GGU Gly GGC Gly GGA Gly GGG Gly	U C A G

Figure 5.9. The genetic code

The genetic code has the following characteristics:

1. The code contains many synonyms; in other words, almost every amino acid is represented by more than one codon.
2. The code has a specific structure. Codons are almost always synonyms and are located in the same "square" (i.e., they differ from each other by the last nucleotide). Exceptions to this rule are the codons for arginine, serine, and leucine, which each have six codons and therefore cannot all fit within a single square that represents all four codons. As can be seen from the genetic code table, the more frequent use of proline, arginine, alanine, and glycine in protein synthesis corresponds to a higher content of [G] + [C], whereas in bacterial proteins with lower [G] + [C] content are expected to contain higher frequencies of phenylalanine, methionine, asparagine, tyrosine, isoleucine, and lysine. Analyses of the amino acid composition of proteins from different bacterial species have confirmed that the varying [G] + [C] contents in their DNA can be partly attributed to these differences in the primary structure of the proteins.
3. The code contains three nonsense codons (UAG, UAA, and UGA), which are marked as "Stop" signals; none of them correspond to any amino acid. These codons serve as signals to terminate the synthesis of polypeptide chains.

The genetic code table allows for a theoretical analysis of the data regarding amino acid substitutions in mutant proteins. This information can be used to test the important hypothesis that mutations resulting in the substitution of one amino acid generally occur as a result of the replacement of identical bases within genetic polynucleotides. However, science has not yet fully elucidated all the mysteries and characteristics of the genetic code.

The main task is to fully decode this code, including explaining the origin of the number 20 amino acids, given that there are 64 triplets. There are also several other questions for which almost no one can provide an answer at the moment.

- Why is the genetic code what it is?
- Why is the genetic code universal?
- Why is the alphabet of the genetic language composed of four letters?
- Why does the genetic code have a triplet structure?
- Why is it precisely this set of amino acids that exists, and not some other combination?
- Why do proteins contain two types of amino acids: polar ones that can form hydrogen bonds, and non-polar ones?
- Why do amino acids have these specific sizes, and not others?
- Why is the code endowed with precisely such a system of degeneracy?
- Why is the genetic code non-overlapping? In other words, why does the cellular translational machinery, which reads the genetic information, use a discrete unit equal to three instead of one?
- Is the emergence of these dialects (the mitochondrial version) accidental or systematic? [104].

A gene controls all the chemical reactions within an organism, thereby determining its characteristics. A unique feature of a gene is its combination of high stability (remaining unchanged across generations) with the ability to undergo hereditary changes; these hereditary changes are the source of genetic diversity among organisms and form the basis for the process of natural selection. "…all life on the planet uses the same genetic code; in other words, life is unified and 'speaks the same language'" [105]. The genetic code is universal because the same codon always corresponds to the same amino acid in all organisms.

The genetic code is a molecular system that facilitates the conversion of information from DNA into proteins. The genetic code is not random in its composition; however, the question of how it came into existence remains unresolved.

The genome is the collection of genes found within the haploid (single) set of chromosomes of an organism. The genome contains the information that determines the hereditary characteristics of a person, which is encoded in the form of genetic material. It has its own "letters," "words," and "grammar rules" (i.e., the rules that govern how genetic information is structured). The human genome consists of tens of thousands of genes. According to some sources, the human genome contains approximately 40,000 genes [2]; other sources estimate that the number of genes in the human genome ranges from 60,000 to 80,000, denoted as [105].

If a codon changes to another codon that corresponds to a different amino acid, the resulting codon has an altered meaning. There are three codons that serve as stop signals, indicating the end of the protein chain; these are called "terminator codons" [106]. It has been observed that the same amino acid can be encoded by multiple three-letter codons. Therefore, the genetic code is somewhat redundant; in genetic terms, this redundancy is referred to as "degeneracy." Each amino acid is encoded by one, two, three, four, or six nucleotide triplets. All codons are continuous and not separated by any special "delimiter" characters (such as commas). The role of this genetic code redundancy is now clear: due to various external influences, errors can occur in the nucleotide chain. However, because of the code's redundancy, not all of these errors result in changes to the synthesized amino acids—or mutations that could be lethal to the organism. In other words, the degeneracy of the genetic code enhances its resistance to interference. Additionally, the code is non-overlapping, meaning reading proceeds codon by codon. All codons are unambiguous, meaning they each encode a single amino acid; consequently, the genetic code from codon to amino acid is also unambiguous. Two codons that encode the same amino acid are considered synonymous.

Let's conduct a more detailed analysis of how triplets are formed using combinatorics. Three groups of triples can be identified when:

1. All three bases are identical (4 groups of triplets): AAA, GGG, TTT, CCC.
2. All three bases are different (4 groups of triplets):

CAT,	TCA,	GCT,	CTG,	TGA,	GAT,	CAG,	AGC
ATC,	ACT,	TGC,	TCG,	ATG,	AGT,	GCA,	GAC
TAC,	CTA,	GTC,	CGT,	TAG,	GTA,	CGA,	ACG

3. Two of the three bases are identical (12 groups of triplets):

TTC,	TTA,	TTG,	CCT,	CCA,	CCG,	AAT,	AAC,	AAG,	GGT,	GGC,	GGA
CTT,	ATT,	GTT,	TCC,	ACC,	GCC,	TAA,	CAA,	GAA,	TGG,	CGG,	AGG
TCT,	TAT,	TGT,	CTC,	CAC,	CGC,	ATA,	ACA,	AGA,	GTG,	GCG,	GAG

From basic combinatorics, it is clear that all triplets, when divided into three families, yield 20 groups. (this is the number of combinations with repetition of any four bases taken three at a time) These 20 combinations correspond to the 20 different amino acids. This might be the answer to the question of why there are only 20 amino acids.

When reading triplets in one step, only one base is changed; the other two bases remain unchanged. Therefore, for three coding bases (one for each amino acid), two bases are repeated in one reading step, while one base is altered. In the next step, another base is changed. A complete change of all bases occurs only after the third step, and at this point, the entire combination of the triplet is no longer affected by previous combinations. This gradual transition from one triplet to another limits the possibility of quickly utilizing all possible permutations of the triplet. [Nikitin A.V., "Problems in Understanding the DNA Coding System."] To utilize all 64 possible triplets, 64 × 3 = 192 individual steps are required for reading the DNA triplets. In order to understand the natural process of amino acid coding in DNA, A.V. Nikitin applies the principle of symmetry in searching for the appropriate triplet combinations. He identifies all symmetric code pairs as follows: 1. Phe – UUU 2. Pro – CCC 3. Lys – AAA 4. Gly – GGG 5. Ile – AUA 6. Arg – AGA 7. Thr – ACA 8. Tyr – UAU 9. Cys – UGU 10. Ser – UCU 11. Glu – GAG 12. Val – GUG 13. Ala – GCG 14. His – CAC 15. Leu – CUC 16. Gln – CAA 17. Asn – AAC 18. Trp – UGG 19. Ter – UAA 20. Asp – GAC 21. Met – AUG 22. Ter – UAG 23. Ter – UGA

Fifteen amino acids were assigned symmetric codons. Afterwards, it appears that nature utilizes two repeated bases within a triplet (CAA, AAC, UGG, UAA) when forming codons. Subsequently, codons consisting of different bases are used as well. It is likely that only the fully symmetric triplets were used from the beginning, while the remaining ones were drawn from an unused pool of possibilities; in this way, all 64 possible triplet combinations were utilized. According to A.V. Nikitin, this allowed the ribosome to perform readout operations in a step-by-step manner, but it also enabled the possibility of triplet reading in an independent mode. Therefore, at that stage, the triplet represents the maximum possible form of information unit in the system: one base on the axis of symmetry, and one base on each of the left and right sides. Three different units of information are used—even for readout processes alone; this is quite complex. In the DNA and RNA coding systems, 4 possible logical states are employed, along with triplet reading. The complexity for the cell is thus extreme. The redundancy in the triplet code is a necessary consequence of the initial method of code reading. This is simply how things evolved [Nikitin A.V., "Problems in Understanding the DNA Coding System"].

Each nucleotide possesses several properties. For example, the two letters A and G are both classified as purines, while the letters T and C are classified as pyrimidines. Based on the presence of weak or strong hydrogen bonds in their complementary pairs, we have: A = T (a complementary pair with two hydrogen bonds) and G ≡ C (a complementary pair with three hydrogen bonds). These properties can be represented using binary numbers: 0 or 1. By assigning to pyrimidines the presence of the trait, i.e. "1", and to purines its absence, i.e. "0", and correspondingly assigning "1" to bases forming complementary pairs via three hydrogen bonds and "0" to those forming complementary pairs via two hydrogen bonds, we adopt the following correspondence between nucleotides and binary numbers:

A = 00; T = 10; G = 01; C = 11.

The linear sequence of these four letters on the nucleic acid strands contains the genetic information necessary for the synthesis of proteins. It is often said that the "encyclopedia of life" is written in these four letters. This set of four letters is considered to be the alphabet of the genetic code.

The genetic code can be associated with the Gray code. The advantage of the Gray code is that it assigns a unique code combination (a binary vector) to each natural number starting from zero; the number of components in this vector depends on the number of numbers being encoded. For example:

1. 0 0 0 0
2. 0 0 0 1
3. 0 0 1 0
4. 0 0 1 1
5. 0 1 0 0
6. 0 1 0 1 etc.

Let us write out all the triplets and their corresponding values (Table 5.1).

Table 5.1

Record of the triplets (codons) and their corresponding amino acids, in order of the increasing binary code values:

Number, triplets	*Binary code:*	*The name of the amino acid:*	*Lat.*	*The literal representation of amino acids...*
1. AAA	00 00 00	Lysine	Lys	K
2. AAG	00 00 01	Lysine	Lys	K
3. AAT	00 00 10	Asparagine	Asn	N
4. AAC	00 00 11	Asparagine	Asn	N
5. AGA	00 01 00	Arginine	Arg	R
6. AGG	00 01 01	Arginine	Arg	R
7. AGT	00 01 10	Serine	Ser	S
8. AGC	00 01 11	Serine	Ser	S
9. ATA	00 10 00	Isoleucine	Ile	I
10. ATG	00 10 01	Methionine	Met	M
11. ATT	00 10 10	Isoleucine	Ile	I
12. ATC	00 10 11	Isoleucine	Ile	I
13. ACA	00 11 00	Threonine	Thr	T
14. ACG	00 11 01	Threonine	Thr	T
15. ACT	00 11 10	Threonine	Thr	T
16. ACC	00 11 11	Threonine	Thr	T
17. GAA	01 00 00	Glutamic acid	Glu	E
18. GAG	01 00 01	Glutamic acid	Glu	E
19. GAT	01 00 10	Aspartic acid	Asp	D
20. GAC	01 00 11	Aspartic acid	Asp	D
21. GGA	01 01 00	Glycine	Gly	G
22. GGG	01 01 01	Glycine	Gly	G
23. GGT	01 01 10	Glycine	Gly	G
24. GGC	01 01 11	Glycine	Gly	G
25. GTA	01 10 00	Valine	Val	V
26. GTG	01 10 01	Valine	Val	V
27. GTT	01 10 10	Valine	Val	V
28. GTC	01 10 11	Valine	Val	V
29. GCA	01 11 00	Alanine	Ala	A
30. GCG	01 11 01	Alanine	Ala	A
31. GCT	01 11 10	Alanine	Ala	A
32. GCC	01 11 11	Alanine	Ala	A
33. TAA	10 00 00	Stop	Term	
34. TAG	10 00 01	Stop	Term	
35. TAT	10 00 10	Tyrosine	Tyr	Y
36. TAC	10 00 11	Tyrosine	Tyr	Y

Number, triplets	*Binary code:*	*The name of the amino acid:*	*Lat.*	*The literal representation of amino acids…*
37. TGA	10 01 00	Stop	Term	
38. TGG	10 01 01	Tryptophan	Trp	W
39. TGT	10 01 10	Cysteine	Cys	C
40. TGC	10 01 11	Cysteine	Cys	C
41. TTA	10 10 00	Leucine	Leu	L
42. TTG	10 10 01	Leucine	Leu	L
43. TTT	10 10 10	Phenylalanine	Phe	F
44. TTC	10 10 11	Phenylalanine	Phe	F
45. TCA	10 11 00	Serine	Ser	S
46. TCG	10 11 01	Serine	Ser	S
47. TCT	10 11 10	Serine	Ser	S
48. TCC	10 11 11	Serine	Ser	S
49. CAA	11 00 00	Glutamine	Gln	Q
50. CAG	11 00 01	Glutamine	Gln	Q
51. CAT	11 00 10	Histidine	His	H
52. CAC	11 00 11	Histidine	His	H
53. CGA	11 01 00	Arginine	Arg	R
54. CGG	11 01 01	Arginine	Arg	R
55. CGT	11 01 10	Arginine	Arg	R
56. CGC	11 01 11	Arginine	Arg	R
57. CTA	11 10 00	Leucine	Leu	L
58. CTG	11 10 01	Leucine	Leu	L
59. CTT	11 10 10	Leucine	Leu	L
60. CTC	11 10 11	Leucine	Leu	L
61. CCA	11 11 00	Proline	Pro	P
62. CCG	11 11 01	Proline	Pro	P
63. CCT	11 11 10	Proline	Pro	P
64. CCC	11 11 11	Proline	Pro	P

In 1961, the American biochemist M. Nirenberg conducted groundbreaking research on deciphering the genetic code. He managed to "read" the first "word" in the DNA sequence: a sequence of three letters—AAA, which corresponded to three adenines arranged one after another [99]. The discovery of the correspondence between codons and amino acids was a major achievement in molecular biology, biochemistry, and biophysics. The implementation of the genetic code within living cells, that is, the synthesis of a protein encoded by a gene, relies on two template-based processes: transcription and translation.

The first process, transcription, takes place in the nucleus and involves the synthesis of molecules of messenger ribonucleic acid (mRNA) at specific regions of DNA. During this process, the nucleotide sequence of DNA is "transcribed" into the corresponding nucleotide sequence of mRNA.

The second process, translation, takes place in the cytoplasm within the cellular organelles responsible for protein synthesis—ribosomes.

Table 5.1 is essentially a codon-amino acid dictionary. The assignment of meanings to codons has raised new questions: Does the genetic code have any physical or molecular significance? Is the correlation between codons and amino acids entirely random? What can be said about the evolution of this code? What factors influence the process of decoding the code, as well as transcription and translation? What mechanisms cause the code to be altered (distorted)? What are the physicochemical reasons behind mutations?

The code represents a translation program. It is of utmost importance to understand the conditions under which this program operates. Only the primary structure of a protein is genetically encoded; however, its biological functions are determined by its spatial structure. The primary structure and the spatial structure are inherently connected to each other. Consequently, both the spatial structure of the protein and its biological functions are genetically encoded. At the same time, natural selection acts not on the primary structure but on the spatial structure—the protein's biological behavior.

As can be seen from Table 5.1, amino acids have the same letters in their first and second positions; however, the third letters vary. For example, the amino acid lysine is encoded by the first base 'A', the second base 'A', and the third base either 'A' or 'G'. Similarly, the triplet sequence encoding the amino acid asparagine has the letter 'A' in the first position, 'A' in the second position, and the letters 'T' or 'C' in the third position. The table also indicates that the degeneracy of the code is uneven. It should be noted that the relationships between nitrogenous bases and binary values were established a long time ago, and the assignments of bases to binary values were done in various ways; for instance, the value 'A' could correspond to either 00 or 10, and the same was true for other bases as well. When referring to sequences in the future, we will mostly be referring to sets of binary numbers, since computers can ultimately process data only in this form.

Genetics suggests that the total number of genetic codes, or their diversity, is estimated to be around 10^{1000}. The selection of a particular genetic code throughout the entire history of the universe (which scientists believe to be 13.5 billion years) occurred in some incredibly "unusual" manner; after all, a complete enumeration of all possible genetic codes seems impossible. Consequently, there must exist other mechanisms or approaches for the generation of genetic codes, such as [20].

5.4. READING FRAMES GENERATED BY CODE

In the mid-20th century, E. Tatum and G. Beadle formulated the principle of "one gene, one enzyme," which became the foundation of biochemical genetics. For this discovery, they were awarded the Nobel Prize in 1958. Since then, it has been believed that each gene codes for one protein. Genes that encode protein sequences are called structural genes.

Since each amino acid is encoded by three nucleotides, one gene corresponds to three reading frames for each DNA chain. For each DNA chain, there is one correct reading frame, which is also called the open reading frame (ORF), as well as two alternative reading frames. The difference between them lies in the following: the ORF contains only one termination codon ("ter") located at the end of the gene, whereas in the alternative reading frames (which are shifted by +1 or –1 nucleotides relative to the ORF), stop codons occur quite frequently. For each of the two DNA chains, the reading frames are interpreted in different directions; it is conventional to read the frames on the first chain from left to right. In the alternative reading frames, in most cases, there is an inhibition of protein synthesis compared to the ORF. This inhibition arises because the shifts of +1 or –1 nucleotides result in a sequence of codons that differs from the main sequence (which does not contain such shifts). In the alternative sequences, one of the three termination codons ("TAA", "TAG", "TGA") appears, causing the cessation of protein synthesis. Thus, nature has ensured a form of biological protection. All amino acids, under a shift of –1 or +1 nucleotide, form corresponding junction pairs (listed in Appendix 1). The total number of amino acids formed and their names are listed in Table 5.2.

It can be inferred from Table 5.2 that for each alternative reading frame of a single chain, up to 80 amino acids can be generated when the reading frame is shifted by -1 or +1 nucleotides.

For example, if we choose the amino acid lysine, whose codons in binary code are either 00 00 00 or 00 00 01, then for the two alternative reading frames, the pattern will look as follows:

	Lys	
–1	00 00 00	
Lys	00 00 00	
Glu	01 00 00	
Stop	10 00 00	
Gln	11 00 00	+1
	00 00 00	Lys
	00 00 01	Lys
	00 00 10	Asn
	00 00 11	Asn

	Lys	
–1	00 00 01	
Lys	00 00 01	
Glu	01 00 01	
Stop	10 00 01	
Gln	11 00 01	+1
	00 01 00	Arg
	00 01 01	Arg
	00 01 10	Ser
	00 01 11	Ser

For each of the twenty amino acids (including all their triplets), at least one alternative amino acid is formed in the alternative frames. Some amino acids—Lys, Gly, Leu, Phe, and Pro—also appear when the nucleotide position is shifted by either –1 or +1 in the alternative frames; in other words, the amino acid itself may not change at all.

The minimum value of the amino acids formed corresponds to tryptophan in the second alternative framework, and it is 1. The maximum value corresponds to leucine in the first alternative framework, and it is 8.

It is possible to identify restrictions (blockages) in the formation of alternative frames for certain amino acids. In this case, when the reading frame of an amino acid from the original open reading frame (ORF) is shifted by –1 or +1 nucleotides, those amino acids will generate codons that indicate the end of the protein sequence within these alternative reading frames.

Table 5.2

Amino acids formed in two alternative reading frames when the ORF is shifted by −1 or +1 nucleotide

Amino acids in the ORF	Amino acids formed in the first frame (shift by −1 nt)	Count	Amino acids formed in the second frame (shift by +1 nt)	Count
Lys	Lys, Glu, Gln	3	Lys, Asn, Arg, Ser	4
Asn	Lys, Glu, Gln	3	Ile, Met, Thr	3
Arg	Lys, Glu, Gln, Thr, Ala, Ser, Pro	7	Glu, Asp, Gly, Val, Ala	5
Ser	Lys, Glu, Gln, Ile, Val, Phe, Leu	7	Val, Ala, Gln, His, Arg, Leu	7
Ile	Asn, Asp, Tyr, His	4	Tyr, Leu, Phe, Ser	4
Met	Asn, Asp, Tyr, His	4	Trp, Tyr	2
Thr	Asn, Asp, Tyr, His	4	Gln, His, Arg, Leu, Pro	5
Glu	Arg, Gly	2	Lys, Asn, Arg, Ser	4
Asp	Arg, Gly	2	Ile, Met, Thr	3
Gly	Arg, Gly, Trp	3	Glu, Asp, Gly, Val, Ala	5
Val	Ser, Gly, Cys, Arg	4	Tyr, Trp, Cys, Leu, Phe, Ser	6
Ala	Ser, Gly, Cys, Arg	4	Gln, His, Arg, Leu, Pro	5
Tyr	Ile, Val, Leu	3	Ile, Met, Thr	3
Trp	Met, Val, Leu	3	Gly	1
Cys	Met, Val, Leu	3	Val, Ala	2
Leu	Ile, Val, Phe, Leu, Thr, Ala, Ser, Pro	8	Tyr, Trp, Cys, Leu, Phe, Ser	6
Phe	Ile, Val, Phe, Leu	4	Leu, Phe, Ser	3
Gln	Thr, Ala, Ser, Pro	4	Lys, Asn, Arg, Ser	4
His	Thr, Ala, Ser, Pro	4	Ile, Met, Thr	3

Amino acids in the ORF	Amino acids formed in the first frame (shift by −1 nt)	Count	Amino acids formed in the second frame (shift by +1 nt)	Count
Pro	Thr, Ala, Ser, Pro	4	Gln, His, Arg, Leu, Pro	5
Σ		**80**		**80**

Of the two DNA strands, one contains information about the structure of a protein (the "sense strand"); the other has no semantic meaning (the "antisense strand"). The sense strand serves as a template for the synthesis of mRNA with the participation of RNA polymerase.

There are special signals in messenger RNA (mRNA) that alter the reading frame. When mRNA contains termination codons in the transcribed region, in order to bypass these codons as "obstacles," signals are sent in that change the reading frame before or directly at those codons. However, it is very difficult to explain such changes in the mRNA reading frame from the perspective of physical or chemical processes.

Let's consider the formation of terminal codons when the phase shifts by either -1 or +1 nucleotides. This information could be useful for further research.

The first reading frame will be the one that has a phase shift of −1 nucleotide. In this case, the formation of the stop codon for the first amino acid in the subsequent pair will involve the third nucleotide: $**h_{13}**$. For the second amino acid, two nucleotides will be involved: one located at the first position of the second amino acid ($*h_{21}$) and the other at the second position ($h_{22}*$).

The lower digit of a nucleotide corresponds to the position of the first or second amino acid in the corresponding pair, while the upper digit corresponds to the position of the nucleotide in the next amino acid pair. Since there are three termination codons (TAA, TAG, and TGA), the numbers of the nucleotides for the first reading frame, as well as the corresponding Boolean variables for the amino acid pairs, must be determined according to the previously established mappings between nucleotides and binary numbers (as described in section 5.2).

For the codon TAA: $x_{13} = 10$ (T); $x_{21} = 00$ (A); $x_{22} = 00$ (A);

For the TAG codon: $x_{13} = 10$ (T); $x_{21} = 00$ (A); $x_{22} = 01$ (G).

For the TGA codon: $x_{13} = 10$ (T); $x_{21} = 01$ (G); $x_{22} = 00$ (A).

For the second reading frame (with a phase shift of +1 nucleotide), the first amino acid in the pair will consist of two nucleotides: x_{12} and x_{13}, while the second amino acid in the pair will consist of only one nucleotide, x_{21}. The nucleotide numbers for the second reading frame, as well as the corresponding Boolean variables in the amino acid pairs, should be adjusted accordingly.

For the codon TAA: $x_{12} = 10$ (A); $x_{13} = 00$ (A); $x_{21} = 00$ (A);

For the TAG codon: $x_{12} = 10$ (A); $x_{13} = 00$ (A); $x_{21} = 01$ (G);

For the TGA codon: $x_{12} = 10$ (A); $x_{13} = 01$ (G); $x_{21} = 00$ (A).

The selection of amino acids for forming pairs is made from Table 5.1. Let's consider all six cases (three for each of the two alternative interpretation frameworks).

We use the method of pairwise comparisons to identify possible combinations of amino acids. In recent times, this comparative approach has been widely and very effectively employed in molecular biology and genetics [3].

Before proceeding to construct pairs of amino acids to determine whether they can form termination codons and to create Table 5.3, let's consider the information from Table 5.1. In Table 5.1, the amino acid number 38 (tryptophan, TGG) has the binary code 10 01 01, while the amino acids glutamine (numbers 49 and 50) have the binary codes 11 00 00 and 11 00 01 (CAA, CAG). These two amino acids cannot possibly form termination codons, either as the first or as the second member of a pair. Now, let's look at the amino acids located in the second half of Table 5.1; starting from position 35 and going all the way to the end of the table, for all of these amino acids, the first nucleotide is either T (10) or C (11). However, these nucleotides cannot form termination codons when they are paired with the second amino acid in a pair, since the second position in such a pair can only have the values G (01) or A (00). Therefore, the second amino acid in the pair cannot be the following amino acids:

1. Tyrosine (all its triplets);
2. Cystine (all its triplets);
3. Leucine (all its triplets);
4. Phenylalanine (all its triplets);
5. Serine – triplets: TCA, TCG, TCT, TCC (i.e., 4 out of 6);
6. Histidine (all triplets);
7. Arginine triplets: CGA, CGG, CGT, CGC (i.e., 4 out of 6 possible combinations).
8. Proline (all triplets).

We will not include these amino acids in Table 5.3 as second members of pairs, in order to avoid cluttering the table.

The table shows all twenty amino acids vertically, and the amino acids that appear in pairs are indicated horizontally. In the cells of the table, the number 1 represents the amino acid pairs that form stop codons for the first reading frame, while the number 2 represents the amino acid pairs that form stop codons for the second reading frame.

Let's calculate the number of amino acid pairs that contribute to the formation of stop codons in the first reading frame. The first amino acids in these pairs are: Asn, Arg, Ser, Thr, Asp, Gly, Ala, Tyr, Cys, Phe, His, Pro (a total of 12 amino acids). The second amino acids in these pairs are: Lys, Asn, Arg, Ser, Glu, Asp (a total of 6 amino acids). Therefore, a total of 72 amino acid pairs form the stop codons for the first reading frame. For the formation of the second reading frame, the first amino acids in the pairs are Ile, Met, Val, Leu. The corresponding second amino acids are as follows: 1. For the first amino acid Ile: 1. Ile; 2. Met; 3. Thr; 4. Gly; 5. Val; 6. Ala. 2. For the first amino acid Met: 1. Lys; 2. Asn; 3. Arg; 4. Ser; 5. Ile; 6. Met; 7. Thr. 3. For the first amino acids Val and Leu: 1. Ile; 2. Thr; 3. Gly; 4. Val; 5. Ala. Thus, a total of 23 amino acid pairs form stop codons for the second terminating reading frame.

In the formation of the second terminating reading frame, 23 pairs of amino acids are involved. Two amino acids—lysine and glutamic acid—do not participate in the formation of terminating stop codons as the first amino acid in a pair. Tryptophan (TGG) and glutamine (CAA, CAG), as already noted, cannot form terminating codons as either the first or the second amino acid in a pair.

Let's count the number of amino acid pairs that are involved in both terminal coding regions simultaneously. For these pairs, the first amino acids in each pair will be: 1. Ile; 2. Val; 3. Leu. The second amino acids in each pair will be: 1. Lys; 2. Asn; 3. Arg; 4. Ser; 5. Glu; 6. Asp. There are a total of 18 such pairs.

Upon conducting a more detailed analysis of the amino acid pairs that form stop codons for both reading frames, it becomes evident that the first amino acids in these frames do differ in their respective triplets. For instance, if the first reading frame is initiated by the amino acid Ile (Isoleucine), which has the triplet code ATT (00 10 10), and the subsequent amino acids are: 1. Lys (Lysine); 2. Asn (Asparagine); 3. Arg (Arginine); 4. Ser (Serine); 5. Glu (Glutamate); 6. Asp (Aspartic Acid), then the first amino acid of the second reading frame is coded by the triplet ATA (00 10 00). Moreover, the triplet codes for both reading frames are identical for the second pair of amino acids. This difference in triplet codes is also observed for the first two amino acids in the second reading frame: Valine (Valine) and Leucine (Leucine). Such differences in triplet codes can be seen for all pairs of amino acids in both reading frames.

As a further step in our analysis, we will consider combinations of three amino acids within a single DNA strand. We will select these third amino acids from the 61 possible meaningful triplets listed in Table 5.1. To simplify the process, we will focus on the second alternative reading frame, as it only contains 23 pairs of amino acids that can form stop codons. It is noteworthy that only the amino acids Ile (Isoleucine), Met (Methionine), and Val (Valine) can potentially form stop codons when paired with a third amino acid in the sequence. For the second reading frame, the second and third nucleotides for these amino acids must be either T = 10 and A = 00 or T = 10 and G = 01; accordingly, these amino acids are Ile, Met, and Val. Following Ile, other possible amino acids could have the first nucleotides A = 00 or G = 01, resulting in the following sequences: Lys (Lysine), Asn, Arg, Ser, Ile, Met, Thr (Threonine), Glu (Glutamate), Asp, Gly (Glycine), Val, Ala (Alanine).

For the amino acid Met (triplet code 001001) and Val (triplet codes 011000, 011001), the possible second amino acids in their pairs are Lys, Asn, Arg, Ser, Ile, Met, Thr. Regardless of the number of amino acid pairs considered for the second reading frame (four, five, etc.), only three amino acids—Ile (triplet 001000), Met (triplet 001001), and Val (triplets 011000, 011001)—can form stop codons when paired with the remaining amino acids. Therefore, if the sequences of amino acid pairs do not contain the aforementioned amino acids Ile, Met, and Val along with their corresponding codons, no stop codons will be generated in the second reading frame, meaning that this frame would not be prohibitive to protein synthesis. This finding supports the possibility of having two or three open reading frames within a single DNA strand.

A complete list of all possible amino acid pairs in the alternative reading frames can be compiled by shifting the phase of the DNA sequence by -1 and +1 nucleotides; these pairs form stop codons within those reading frames.

Yes, the amino acids Lys (positions 1, 2), Asn (positions 3, 4), Arg (positions 5, 6), Ser (positions 7, 8), Glu (positions 17, 18), Asp (positions 19, 20) form stop codons when the nucleotide sequence is shifted by –1. The same applies to the second alternative reading frame, i.e., when the nucleotide sequence is shifted by +1.

In the second reading frame, the stop codons encode the amino acids Ile (position 9), Met (position 10), Val (positions 25, 26), and Leu (positions 41, 42, 57, 58).

In the list of codon–amino acid pairs (see Appendix 1), pairs that form terminating stop codons in the alternative reading frames are marked with a bold dot to the right of the corresponding number. Let us trace the formation of terminating stop codons using the fragment of genome SV40 shown in Figure 5.11 as an example. As can be seen from Figure 5.11, the terminating codon in the first alternative reading frame (shift –1) is formed by Lysine (AAA) from the ORF. For the second alternative reading frame (shift +1), the terminating stop codon is formed by Leucine (TTA) from the ORF.

For example:

```
                GATGGATAAAGTTTTAAACAGAGAGGAATCTTTG
                                •
ORF        Met  Asp  Lys  Val  Leu  Asn  Arg  Glu  Glu  Ser  Leu
Shift -1   Asp  Gly  Stop Ser  Phe  Lys  Gln  Arg  Gly  Ile  Phe
                     ----
Shift +1   Trp  Ile  Lys  Phe  Stop Thr  Glu  Arg  Asn  Leu
                               ----
```

Figure 5.11. A portion of the SV40 genome fragment, demonstrating all three reading frames for that single DNA chain

Table 5.3

	Lys	Asn	Arg	Ser	Ile	Met	Thr	Glu	Asp	Gly	Val
	1	2	3	4	5	6	7	8	9	10	11
	1	1	1	1				1	1		
	1	1	1	1				1	1		
	1	1	1	1				1	1		
	1	1	1	1				1	1		
	1	1	1	1				1	1		
	2	2	2	2	2	2	2	2	2	2	2
	2	2	2	2	2	2	2				
	1	1	1	1				1	1		
	1	1	1	1				1	1		
	1	1	1	1				1	1		
	1	1	1	1				1	1		
	2	2	2	2	2	2	2	2	2	2	2
	2	2	2	2	2	2	2				
	1	1	1	1				1	1		
	1	1	1	1				1	1		

	Lys	Asn	Arg	Ser	Ile	Met	Thr	Glu	Asp	Gly
TGT	1	1	1	1				1	1	
CTT	1	1	1	1				1	1	
TTA	2	2	2	2	2	2	2	2	2	2
CTA	2	2	2	2	2	2	2	2	2	2
TTG	2	2	2	2	2	2	2			
CTG	2	2	2	2	2	2	2			
TTT	1	1	1	1				1	1	
Gln										
CAT	1	1	1	1				1	1	
CCT	1	1	1	1				1	1	

In alternative reading frames, it is possible to predict the formation of not only stop codons but also amino acids based on the list of codon pairs. For example, consider the amino acid Arg (AGA) in the ORF (Open Reading Frame); in the ORF, it is indicated with a dot at the top. In the list of codon pairs (Appendix 1), it corresponds to position 5. For the first reading frame, it can form the amino acids Lys, Glu, and Gln. In our example shown in Figure 5.11, it forms the amino acid Gln, while for the second reading frame it forms the amino acid Glu, which is also within its coding possibilities (in addition to Glu, it can also encode the amino acid Asp).

In recent times, computer methods have been actively used for the analysis and recognition of genetic information. A new field of informatics is emerging: bioinformatics.

Bioinformatic methods are becoming increasingly important as the number of sequenced genomes increases. They enable the identification of genes within DNA sequences, the prediction of proteins encoded by these genes, the detection of conserved domains in proteins, the prediction of the biochemical and physical properties of proteins, and the comparative analysis of genes and their products across different organisms [107].

To identify the function of an ORF (Open Reading Frame) that is detected during the decoding of a genomic sequence, a search is first performed for homologous genes in other organisms. These homologous genes share similarities in their sequence composition and may have a common origin. The search is carried out in databases that contain gene and genome sequences using specialized programs such as BLAST (Basic Local Alignment Search Tool; http://www.ncbi.nlm.nih.gov/blast/).

Sometimes, homology searches are performed between DNA sequences. However, a high degree of similarity in nucleotide sequences does not guarantee a high degree of similarity in the corresponding amino acid sequences (Figure 5.12, A). Therefore, DNA is usually "translated" into an amino acid sequence. For prokaryotes, this translation involves translating nucleotide triplets into their respective amino acids. For eukaryotes, the first step is to identify the coding sequences (exons) in the mRNA; these exons are identified based on the presence of conserved splicing sites. Subsequently, homologs are searched for among the amino acid sequences stored in databases. The analysis of amino acid sequence homology is the most informative, as it can indicate similarities in protein function. It should be noted that mutations occur in DNA more frequently than changes in amino acid sequences. Thus, two homologous genes may have highly similar amino acid sequences despite only a weak similarity in their nucleotide sequences; this is due to the redundancy of the genetic code (one amino acid can be encoded by multiple codons) (Figure 5.12B). [107].

A

Identity: By nucleotides (|) – 77%; by amino acids (•) – 30%.

```
 G  A  P  G  M  W  L  R  L  A
GGTGCACCCGGTATGTGACTGCGATTAGCA
| | ||||| |||| |||| || ||| |||
GATACACCCCGTATTTGACAGCAATTTGCA
 D  T  P  R  I  W  E  E  F  A
       •        •           •
```

B

Identity: By nucleotides (|) – 37%; by amino acids (•) – 80%.

```
 E  K  S  Q  Y  L  T  R  S  A
GAAAAATCTCAATATCTTACTCGCAGTGCT
||       ||   | | ||  |    ||
GAGTCTAGCCAGGGTTTAACAAGATCGGCG
 E  S  S  Q  G  L  T  R  S  A
 •     •  •     •  •  •  •  •
```

Figure 5.12. Comparison of the similarity between two sequences based on their nucleotide and amino acid compositions

The homology of protein sequences can indicate that the proteins perform identical or similar functions. Homologous sequences may exhibit similarities not over their entire length, but only in certain regions (domains). Such regions are referred to as conserved domains. Conserved domains can include catalytic domains, DNA-binding domains, and others.

CHAPTER 6. THE ROLE OF MATHEMATICS IN GENETIC RESEARCH

The German philosopher Immanuel Kant noted: "Every natural science contains as much truth as the amount of mathematics that is incorporated into it." The Italian scientist Galileo Galilei (1564–1642), one of the founders of modern empirical science, stated that the book of Nature is written in the language of mathematics.

Mathematics has successfully tackled the problems that life has posed to it. However, the development of mathematics has not always been smooth; there have been periods of decline as well as periods of growth in mathematical thought. The 20th century witnessed a crisis in the foundations of mathematics, which accelerated the study of logic. In fact, the root cause of this crisis was the widespread emergence of combinatorial enumeration (exhaustive-search) problems. The crisis in mathematics is not surprising, as we can observe similar crises in other fields of science today.

As mentioned in Chapter 1, in the 1930s, Kurt Gödel's incompleteness theorem was published. The essence of this theorem is that in any sufficiently complex theory, there exist statements that cannot be proven or disproven solely using the axioms of that theory; alternatively, it is possible to prove any statement, regardless of whether it is true or false. Gödel's theorem allowed us to view all developments in science from the perspective of "lifting restrictions" on various previously unassailable assertions. It provided a means to understand the dynamic evolution of mathematics (represented by [75]).

Mathematics has played a special role in the process of genetic research. A great contribution by G. Mendel was his ability to find an incredibly simple way to express the types of inherited traits observed in pea crossings, as well as their numerical relationships, in mathematical formulas.

The work of S. Wright, J.B.S. Haldane, and R. Fisher in the 1920s and 1930s laid the foundations for the genetic-mathematical methods used to study processes occurring within populations. In a series of studies, the English biologist J.B.S. Haldane pioneered a new field: mathematical genetics, by investigating how the frequency of genes in populations changes under natural and artificial selection. He wrote: "Science almost always has a quantitative expression; it deals with measurements, weight indicators, and counting." [109]

However, the conclusions drawn by theorists have shown that the direct approach leads to completely hopeless (i.e., unsolvable) equations. If the situation requires the results of theoretical calculations to be of vital importance—for example, in determining the three-dimensional structure of a protein—then approximate methods must be employed. For many years, genetic research has been conducted using mathematical statistics.

This approach was purely empirical and did not address the underlying complexities of genetic issues. The further development of experiments in the field of genetic research was associated with the search for new methods that would enable a shift from quantitative to qualitative analysis.

Mathematics is not directly present in living nature, molecular biology, or genetics. There are no clear formulas or equations that can describe them. Living organisms continuously receive information from the external environment and process it in various ways, resulting in numerous reactions within the organism.

Depending on the concentration of certain regulatory proteins, genes responsible for producing the required proteins are switched. In addition to voluntary (i.e., consciously controlled) reactions, emotions play a significant role; they are accompanied by a wide range of involuntary (automated) physiological responses. Unvoluntary physiological reactions triggered by emotional experiences include changes in the cardiovascular system (pulse rate, blood pressure, blood vessel diameter), respiratory system (breathing rate, depth, and rhythm), body temperature, as well as activity of sweat glands, salivary glands, and other endocrine organs (such as the pituitary gland, thyroid gland, and pancreas), among others.

The organism is a system that undergoes periodic fluctuations. Without exception, all the functions of the organism are governed by the law of rhythm. Rhythm is a form of movement in living matter. It is impossible to predict the true conformation of a protein in advance through analytical methods; likewise, it is not possible to forecast the state of the entire organism at any given moment. Moreover, chemical reactions within the cell occur extremely rapidly. A single enzyme molecule can catalyze thousands of reactions per second. Such high rates are only possible because the molecules move very quickly.

T. Morgan wrote that "an exclusively mathematical approach to the principles of gene segregation and independent assortment would hardly satisfy botanists and zoologists for a long time." [110]

The fundamental starting point of any theory or science is, as we know, the set of principles upon which it is based. Formal theories can be divided into two categories: those with a developing system of principles, and those with a stable system of principles. Theories in the first category continually enrich themselves with new assertions throughout their development; however, the justification for these new assertions cannot be derived from the principles that were previously accepted. A classic example of such a theory in biology is the theory of cell structure. Mathematical theories, on the other hand, belong to the second category. This is precisely why mathematics has limited capabilities when it comes to explaining and understanding the developments in genetics. In mathematics, there are theories whose fundamental principles remain unchanged over time; if they do change, this is considered a transition to a new theory. For instance, Einstein's theory of relativity can be seen as a refinement of Newton's classical mechanics. A stable system of principles is always embodied in axioms. Formal theories, therefore, represent models of scientific theories that are based on such stable sets of principles. This understanding of the place of formal theories in science guards against errors.

Nevertheless, it is precisely the combination of the complex internal structure of mathematics with real-world application problems that drives this science forward. I.P. Pavlov wrote: "All life—from the simplest to the most complex organisms, including humans—represents a long series of increasingly sophisticated mechanisms for balancing external environments. There will come a time (perhaps one far in the future) when mathematical analysis, based on natural sciences, will encompass all these equilibrium processes in magnificent equations of equations, including, ultimately, itself." [111] Pavlov's statements provide grounds for arguing in favor of a deterministic origin of life and could compete with theories of random evolutionary selection. It follows that the functional proteins resulting from biochemical reactions, as well as the self-reproduction of living systems, are determined not only by physical and chemical properties but also by the laws of abstract sciences such as mathematics and logic.

It is precisely through the process of moving from the simple to the complex that mathematics has developed. It began to split into various branches, quickly gaining the status of independent sciences. Over the past few decades, we have witnessed the integration of different scientific disciplines. A new field has emerged: computational intelligence (CI). This field encompasses hybrid systems (genetic-neural, fuzzy-genetic, logical-genetic), knowledge-based approximate reasoning systems, and data-driven systems (neural networks, genetic algorithms). These developments enable more in-depth research in the field of genetics. For example, the Soviet mathematician S.N. Bernstein, who was an academician of the USSR Academy of Sciences (1929) and the Ukrainian SSR Academy of Sciences (1925) and worked at the Kharkov University, tackled genetic problems that required methods from probability theory. The so-called Bernstein algebras or stationary algebras that he developed were not applied in the field of mathematical genetics until 50 years later [112].

The stochastic processes in the functioning of chromosomal DNA are reduced to an optimum. Scientists believe that there is a compromise between randomness and determinism; in other words, chance and order are in balance. In his research, V.I. Shcherbak notes that numbers (along with letters) have become an integral part of the genetic (protein) code. V.I. Shcherbak writes: "Certain cellular organelles must function as biocomputers... and we need to discover the numerical systems they use." He further states: "It seems that the genetic code is more closely related to abstract concepts of arithmetic than to concepts from physics and chemistry." He believes that the binary logic of digital computing in the genome is a decisive factor in its functioning. [113]

Probability theory for a long time remained a mathematical discipline in which the fundamental concepts were not sufficiently clearly defined. This ambiguity often led to paradoxical conclusions (such as Bertrand's Paradox). Naturally, the application of probability theory to the study of natural phenomena was poorly justified and sometimes met with severe criticism. The development of the natural sciences at the beginning of the last century placed higher demands on probability theory. Formal logical justification and an axiomatic construction of the theory became of great importance; probability theory had to be built upon axioms, just like any established mathematical discipline—geometry, theoretical mechanics, abstract group theory, etc. For the first time, this perspective was articulated and developed in 1917 by the Soviet mathematician S.N. Bernstein. A.N. Kolmogorov proposed a different approach, linking probability theory with metric theory of functions as well as set theory. The axiomatic definition of probability incorporates both the classical and statistical interpretations of the concept, overcoming the shortcomings of each. On this basis, it was possible to construct a logically sound foundation for modern probability theory that also meets the higher requirements of contemporary natural science. [114]

Thanks to the efforts of A.N. Kolmogorov [115], the theory of probability is based on four axioms, which are used to introduce the concept of probability and certain properties of sets—both for finite sets of elementary events and for any infinite set. Kolmogorov's axioms are consistent because there exist real objects that satisfy all of Kolmogorov's axioms simultaneously. According to K. Gödel's theory, this consistency must be proven using criteria of a higher level of rigor than those used in probability theory. Therefore, this system of axioms from A.N. Kolmogorov is not complete. However, this lack of completeness allows for flexibility in adapting to specific application scenarios, as it enables the consideration of the same sets of random events with different probabilities in various tasks.

Thus, the inductive approach proposed in work [116] contradicts the very foundation of probability theory, which demonstrates the correctness of an approach based on a mixed inductive-deductive methodology.

6.1. STATISTICAL ANALYSIS OF GENETIC INFORMATION

After the genetic code was deciphered, a new phase of its study began, focusing on the investigation of the systematic properties of the genetic code: symmetry, regularity, universality, etc., all based on statistical analysis.

All protein sequences consist of sets of amino acids, with the number of amino acids being equal to 20.

Each of these amino acids is encoded by a certain number of triplets:
by one triplet — tryptophan, methionine;
by two triplets — lysine, glutamine, histidine, phenylalanine, cystine, tyrosine, asparagine, glutamic acid, aspartic acid;
by three triplets — isoleucine;
by four triplets — glycine, threonine, valine, alanine, proline;
by six triplets — serine, leucine, arginine.

Three of the 64 triplets do not code for any amino acids; they signal the end of protein synthesis and are called stop codons ("ter" codons). These codons are TAA, TAG, and TGA.

As we have already mentioned, a portion of the amino acids is encoded by four triplets; in this case, the third base position is fully degenerate (i.e., any nucleotide at the third position encodes the same amino acid). Another portion of the amino acids is encoded by two triplets, where the third position in one triplet contains either C or T (for one amino acid) and G or A (for another amino acid). This suggests that the first two bases in a triplet are the most decisive factors in determining the identity of the amino acid, while the third base has less influence and provides less information. Based on this, we will disregard the third base and construct Table 6.1. As can be seen from Table 6.1, along one of the diagonals from the top left corner to the bottom right, there are duplets AA–GG, followed by their complementary pairs TT–CC. On either side of this diagonal are other duplets: GA and AG, TA and AT, CA and AC, TG and GT, CG and GC, CT and TC, which are formed by replacing the first letter in each triplet with the second letter.

When analyzing genetic code, certain patterns can be identified, and the following characteristics can be noted:

- Triplets that have purine bases A and G in the second position encode polar amino acids: Lysine (Lys), Asparagine (Asn), Arginine (Arg), Glutamic Acid (Glu), Aspartic Acid (Asp), Glycine (Gly), Tyrosine (Tyr), Tryptophan (Trp), Cysteine (Cys), Glutamine (Gln), Histidine (His), and Serine (Ser).
- Triplet sequences that contain pyrimidine bases T and C in the second position primarily encode non-polar amino acids (Phenylalanine, Leucine,

Isoleucine, Valine, Alanine, Proline, Methionine) as well as weakly polar amino acids (Serine, Threonine).

Table 6.1

AA (Lys, Asn)	AG (Arg, Ser)	AT (Ile, Met)	AC (Thr)
GA (Glu, Asp)	GG (Gly)	GT (Val)	GC (Ala)
TA (Stop, Tyr)	TG (Stop, Trp, Cys)	TT (Leu, Phe)	TC (Ser)
CA (Gln, His)	CG (Arg)	CT (Leu)	CC (Pro)

The analysis indicates that the amino acid serine (Ser) falls within both the category of polar and the category of weakly polar amino acids. This is due to the fact that the side chain of serine is located at the lower limit that defines the size of an amino acid as a bonding entity (represented by [104]). Some authors classify glycine (Gly) as a non-polar amino acid, while threonine (Thr) as a polar one [117]. Tryptophan is classified as a non-polar amino acid by T.T. Berezov and B.F. Korovkin [118]. The classification of cystine (Cys) varies among different authors; some consider it polar, while others classify it as non-polar ([102, 27]).

In Table 6.1, two groups can be identified. The first group (the cells with dark background) contains eight pairs of nucleotides, each encoding one amino acid. The second group (the cells with light background), also consisting of eight pairs of nucleotides, contains pairs that encode two amino acids each.

Furthermore, if the bases in the first group are replaced according to the G–T and C–A rules, then the entire group is completely transformed into the second group. This transformation is known as "Rumer's rule" [119]. These two groups form rotationally symmetrical regions, separated by a broken line; they transform into each other according to Rumera's rule. For example, the CC pair located in the lower right corner is transformed into the AA pair located in the upper left corner.

The genetic code can be represented in Table 6.2 as follows: In Column 1, there are brief abbreviations for the amino acids; in Column 2, their alphabetical names are listed; in Column 3, the corresponding triplets of nucleotides that code for these amino acids are provided; and in Column 4, the number of occurrences of each triplet is indicated.

We know that DNA fragments contain information about the sequence of amino acids in proteins. This sequence determines the formation of the internal molecular bonds within the protein. At one end of a protein chain molecule, there is a NH2 group (an amino group), and at the other end, there is a COOH group (a carboxyl group). This indicates that the sequence of amino acid residues in proteins is directional; in other words, the molecules of any pair of amino acids can be connected to each other in two ways. In the first type of connection, a residue from one amino acid participates in the formation of a peptide bond with its own amino acid, while in the second type of connection, it binds with the carboxyl group of another amino acid.

Table 6.2

	1	2	3	4	
1	Trp	W	ATG	1	
2	Met	M	TGG	1	
3	Lys	K	AAX	2	X: A, G
4	Gln	Q	CAX	2	
5	Glu	E	GAX	2	
6	Phe	F	TTY	2	Y: T, C
7	Tyr	Y	TAY	2	
8	His	H	CAY	2	
9	Asn	N	AAY	2	
10	Asp	D	GAY	2	
11	Cys	C	TGY	3	M: A, T, C
12	Ile	I	ATM	4	
13	Val	V	GTN	4	N: T, C, A, G
14	Pro	P	CCN	4	
15	Thr	T	ACN	4	
16	Ala	A	GCN	4	
17	Gly	G	GGN	4	
18	Ser	S	TCN, AGY	6	
19	Leu	L	CTN, TTX	6	
20	Arg	R	CGN, AGX	6	
	ter *		TAX, TGA	3	

When analyzing the classification of amino acids into different bond groups, we observe that substances like lysine and arginine possess distinct basic properties. Lysine has four CH2 (methylene) groups, and at the end of its structure, there is a C-NH2 group. Lysine typically forms pairs with arginine. Both lysine and arginine belong to the polar amino acid category and are capable of forming hydrogen bonds. They also have the longest chains among all amino acids. Statistics show that lysine and arginine are particularly prevalent in alpha-helical protein segments.

Glutamic acid and aspartic acid are examples of polar amino acids. The side chains of these amino acids in proteins can form hydrogen bonds both with the main chain of the protein and with other side chains of other amino acids.

Proline cannot form hydrogen bonds under any circumstances; in fact, this amino acid disrupts the helical alpha structures.

Alanine can interfere with the area where a hydrogen bond might form, thereby affecting the likelihood of that hydrogen bond being established.

Leucine and the amino acid valine, which have similar physical properties, are more frequently found in beta-structures.

Isoleucine is also frequently found in beta-structures.

Phenylalanine forms anti-bonding operators.

Although methionine is classified as a non-polar amino acid, the sulfur atom present in its structure is capable of forming hydrogen bonds.

Two amino acids – serine and a closely related analog, threonine – belong to the category of weakly polar amino acids. The side chains of these amino acids form hydrogen bonds with the solvent (water), as well as with the main chain and other side chains of the protein. Serine is also included in the group of polar amino acids due to the presence of polar groups in its side chain. Serine and threonine have the shortest side chains bearing polar groups, so hydrogen bond formation for these amino acids is at the limit (marginal). These two amino acids are frequently found in the beta-sheet structural regions of proteins.

Glutamic acid and aspartic acid are both acidic substances.

Phenylalanine, tyrosine, tryptophan, histidine are all compounds that contain aromatic side chains.

There are two groups of amino acids that have been identified: polar amino acids, which are capable of forming hydrogen bonds; and nonpolar or weakly polar amino acids, which generally cannot form hydrogen bonds.

Let's take a closer look at what polar bonds and polar molecules are. To do this, let's refer to the definition of electronegativity: [46]. Electronegativity is a relative measure of an atom's ability to attract electrons within a molecule, thereby acquiring a negative charge. By knowing the electronegativity of atoms, we can determine the polarity of covalent bonds, calculate the effective charges of the atoms, and more. For example, in molecules like HCl or CO, one of the atoms attracts electrons much stronger than the other; therefore, that atom is more electronegative. In hydrogen chloride (HCl), the more electronegative atom is chlorine, and in carbon monoxide (CO), it's oxygen. As a result, the electrons forming the chemical bond shift towards one of the nuclei, creating a negative charge on that nucleus and a positive charge on the other. Such a chemical bond (and the entire molecule) is considered polar, as it possesses a non-zero electric dipole moment. In large protein molecules, there are both polar and non-polar bonds, as well as polar and non-polar groups formed by these bonds. For instance, the carbon-carbon bond (—C—C—) is a non-polar bond, since the atoms have similar electronegativities. The carbon-hydrogen bond (—C—H—) is likewise non-polar for the same reason. However, the following bonds are polar: —C—O—, —C=O, —O—H, —N—H, —C—N—, —C=N—.

The 19th-century German philosopher Arthur Schopenhauer wrote: "Polarity, or the division of a force into two equal and opposite halves, is the fundamental principle underlying all natural phenomena—from crystals and magnets to human beings themselves."

Proteins perform all the vital functions in an organism. Proteins are enzymes that can accelerate various chemical reactions. It is already known that enzyme proteins use different mechanisms to speed up these reactions (for more details, see [120]). Starting from their initial location within the organism's organs, they control all life processes. For example, a person's hands, legs, eyes, and lungs are symmetrical, while the heart and liver are not. The arrangement of the organism's body parts (from right to left) is determined by specific genes; these genes produce proteins. In which direction the proteins move determines the position of the corresponding organs. If the instructions from DNA indicate that the spleen should form on the left, then, upon receiving the signal that spleen formation has begun, the activity of the "right" growth factors is blocked (see [72]). Proteins transport oxygen and store it, ensuring respiration. They are essential for both internal movements within the organism and the organism's overall movement. They also protect the body from diseases. In short, proteins can do everything that an organism needs… but they cannot synthesize themselves. The synthesis of proteins requires other biopolymers called nucleic acids. The combination of amino acids to form a protein chain occurs through the NH2 and COOH groups, with water being released in the process. Amino acid residues are only connected by the peptide bond CO–NH.

The variety of chemical elements in amino acids is limited: carbon, hydrogen, nitrogen, oxygen, and in two amino acids, sulfur.

However, the diversity of the structures and properties of the amino acids that make up proteins is considerable. This diversity is of great importance for the structure and properties of proteins. As indicated in Table 5.1, Codon #10 is ATG, which corresponds to the amino acid methionine (Met), and codon #26 is GTG, which corresponds to the amino acid valine (Val); these are special triplets. They encode the specified amino acids in the middle of the protein chain; generally speaking, however, their role is to initiate the process of building the chain – that is, to start the synthesis of the sequence [121]. Furthermore, it has been discovered that in some cases, the first methionine in a protein is not encoded by the methionine codon ATG, but rather by the codon GTG (valine). In eukaryotes, initiation of the process usually occurs with the first AUG codon. However, this can only happen if the AUG codon is in an optimal context: there must be a purine nucleotide (A or G) located two nucleotides prior to it, and a G nucleotide immediately following it. If the first AUG codon in the eukaryotic mRNA is not in an optimal binding context, it is skipped, and initiation of translation begins with the next AUG codon. Such an initialization was named "cap-dependent initialization" due to the scanning mechanism [122].

Codons: #33 (TAA), #34 (TAG), and #37 (TGA) do not encode any amino acids; they are stop codons.

As we have already mentioned, the analysis of the genetic code indicates that although the code is composed of triplets, the primary semantic burden is carried by the pair of nucleotides at the beginning of a codon. In other words, the code can be considered "quasi-duplet-based." However, there may not be 20 possible combinations of nucleotides in a codon; rather, there are only 16. Therefore, the third nucleotide in a codon must also carry some semantic information. There is a rule that the genetic code follows almost strictly: nucleotides can be divided into two distinct classes based on their structure—purines (A and G) and pyrimidines (T and C). This rule states that if two codons have the same first two nucleotides and their third nucleotides belong to the same class, they encode the same amino acid. This rule is generally observed; however, there are two exceptions. If the rule were followed strictly, codon ATA should correspond to methionine rather than isoleucine, and codon TGA should correspond to tryptophan rather than serve as a stop signal. Due to this "codon redundancy," most of the 20 amino acids are encoded by more than one codon. Thus, some amino acids have multiple codons; such codons are called "synonymous." They usually differ only in their third position. For example, all four codons for valine encode the same amino acid, although GTT is used approximately twice as frequently as GTC. For arginine, the codons CGT and CGC are most commonly used in most organisms, while CGA, CGG, AGA, and AGG are rarely utilized. The codon AAC is used much more frequently for encoding asparagine than AAT. In humans, the codons AGA and AGG are the most frequently used for arginine; E. coli, on the other hand, uses the codon AGA very rarely and often makes errors during translation. The reason for this behavior is that cells select one synonymous codon among all possible ones because doing so increases the speed or accuracy of synthesis of the corresponding protein. It has also been found that tRNAs corresponding to synonymous codons are present in cells in varying amounts. Further statistics have shown that related organisms tend to have similar preferences for certain codons. The phenomenon where a particular codon is used more frequently than others among a set of synonyms is called "codon bias."

Codon preference is related to the number of tRNAs corresponding to those codons, as well as to the degree of expression of the given gene. Highly expressed genes show a much sharper preference for certain codons compared to genes with low expression. Thus, synonymous codons are subject to selective pressure — the use of the most frequent codons increases the accuracy and speed of transcription. A gene that contains more codons corresponding to a tRNA that is in relative excess should be translated faster than others, in order to make use of this excess tRNA supply. There are also other known cases: one type of tRNA that corresponds to several synonymous codons may bind preferentially to the one that ensures maximum accuracy of translation. Consequently, the cell has sufficient grounds to use synonymous codons at different frequencies. Many tRNAs are able to recognize not only their "own" codon but also some others; this phenomenon is called wobbling. As a rule, tRNAs recognize synonymous codons with differing efficiency. Mutations in DNA that replace a codon with a synonymous one do not change the amino acid sequence of the corresponding protein. Such mutations are called silent mutations [123].

Replacing the third nitrogenous base in a triplet almost never has any consequences. Such mutations are also referred to as "silent substitutions."

The polarity of amino acid residues plays a crucial role in the formation of the protein's tertiary structure. Non-polar amino acids form the core of the protein globule, which avoids contact with water molecules. Polar residues, on the other hand, form the outer layer that interacts with the surrounding water molecules. By placing the amino acid residues from the protein's primary structure at the nodes of a three-dimensional lattice whose edge lengths correspond to the length of peptide bonds, we can determine the number of contacts between non-polar residues (NN contacts). The goal is to find a configuration that maximizes the total number of NN contacts. The energy associated with this configuration is represented in a specific mathematical form...

$$E = -n_{HH}(\varpi), \varpi \in \Omega \qquad (6.1)$$

where $n_{hh}(\varpi)$ represents the number of HH-contacts in the configuration ϖ.

Ω represents all possible configurations of the protein with a given primary structure.

To minimize energy consumption, the entire arsenal of modern optimization methods is typically used.

The hydrophobic effect is usually modeled as a function that is proportional to the surface area of the protein's contact with the solvent. However, there are several phenomena observed in real proteins for which this model does not hold true. For example, when different layers of the protein are separated by only a monomolecular layer of solvent, the protein atoms on either side of the solvent layer can interact with each other.

The existing models for accounting for such effects are inaccurate, resource-intensive, and furthermore require additional approximation and algorithmic optimization. It is assumed that it is impossible to establish a procedure for recognizing the structure of a protein based solely on the internal molecular interactions between its atoms. To predict the protein's structure more accurately, it is necessary to take into account the interaction between the protein and its solvent. However, the hydrophobic effect, which causes certain amino acids in the protein to tend to position themselves closer to the protein's core, plays a secondary role (indicated by [10]).

Fischer conducted a rough estimation of the influence of hydrophobic interactions on the shape of the globule [134]. He divided all the amino acid residues into two groups: polar (or hydrophilic) residues (Arg, Asp, His, Glu, Lys, Ser, Tyr, Thr) and non-polar (or hydrophobic) residues (the remaining 12). He assumed that all amino acid residues have approximately the same volume. Hydrophobic interactions cause the hydrophobic residues to be located inside the globule, while the hydrophilic residues are positioned on its surface. The approximate shape of the globule can be determined through simple calculations. Fischer believes that the outer layer of the globule, which consists of hydrophilic residues, is monomolecular in nature and has a constant thickness of d. If the globule is spherical, then the volume of this outer layer is equal to…

$$V_e = \frac{4\pi}{3}[r^3 - (r - d)^3] \qquad (6.2)$$

where r is the radius of the globule.

The internal volume of the globule, that is, the volume of the hydrophobic residues:

$$V_i = \frac{4\pi}{3}(r - d)^3. \qquad (6.3)$$

The ratio of the number of hydrophilic residues to the number of hydrophobic residues:

$b = V_e/V_i$.

This applies to the spherical globe.

$$b = b_s = \frac{r^3}{(r-d)^3} - 1, \qquad (6.4)$$

$$V_e = V_t\,\frac{b}{b+1}, \qquad (6.5)$$

where $V_t = V_e + V_i$ represents the total volume of the globule. Additionally, $V_e \approx Ad$...

where A represents the surface of the hydrophobic core of the globule. Fischer believes...

d ≈ 4Å. Therefore

$$b = \frac{A}{\frac{1}{4}V - A} \quad (6.6)$$

(A is measured in Å, and V is measured in Å3).

The smaller the value of V_t, the lower the molecular weight of the protein will be, and consequently, the higher its relative hydrophilicity (polarity) should be.

For a sphere with radius r → d, the value of $b_s \to \infty$. The globe can be spherical only when $b = b_s$. Typically, $b > b_s$, and in this case the globe takes on the shape of an ellipsoid, which no longer satisfies the condition of having the minimum surface area for a given volume.

When $b > b_s$, the number of hydrophilic residues exceeds the minimum required to protect the hydrophobic core from the aqueous environment. If $b \gg b_s$, a fibrous structure is formed instead of a globular one. Conversely, when $b < b_s$, the hydrophilic residues do not cover the hydrophobic regions properly. As a result, hydrophobic residues that are not protected from the solvent are also present on the surface of the globule. The hydrophobic interactions between these residues can cause the globules to aggregate.

It is evident that non-polar hydrophobic residues can also be present on the surface of the protein globule, as long as they are adjacent to hydrophilic residues in the chain. On the other hand, one might assume that the core of the globule should primarily be hydrophobic, as this would stabilize its structure.

Amino acids can be classified according to their level of hydrophobicity as high, medium, or low. Amino acids with high hydrophobicity include phenylalanine, leucine, isoleucine, methionine, valine, and cystine. Amino acids with medium hydrophobicity include serine, proline, threonine, alanine, tyrosine, glycine, and tryptophan. Amino acids with low hydrophobicity include histidine, glutamine, asparagine, lysine, aspartic acid, glutamic acid, and arginine.

The separation of amino acid residues into hydrophobic and hydrophilic categories is to some extent arbitrary. In fact, what should be discussed is the degree of hydrophobicity of each residue, along with a quantitative measure of this property. Taking into account the degree of hydrophobicity of amino acid residues provides information about the stability of the protein's globular macromolecule through the interactions between its hydrophobic parts. However, such assessments are not sufficient; it is also necessary to consider the actual structure of the protein globule and the overall balance of all the interactions that occur within it. The simplified theory proposed by Fischer does not account for these factors. Brants developed and refined Fischer's idea ([124]).

He considered three types of residues: hydrophobic residues, located predominantly inside the globule (Val, Leu, Phe, etc.); hydrophilic residues, located predominantly on the surface (in particular, all charged residues); and neutral residues (Gly, Ser, Cys, and possibly Asn, Gln, Tyr, and Thr), which are found both inside and outside the globule. The polar charged residues include Arg, His, Lys, which are positively charged, and Glu, Asp, which are negatively charged amino acids.

Studies of the gene structures in humans and various organisms have revealed the presence of modifications and various types of rearrangements in the genetic material. These rearrangements and alterations occur in the non-coding regions of the genome. In contrast, the protein-coding genes are highly stable and of ancient origin; they evolve more slowly than the surrounding non-coding regions. Such genomic rearrangements must not disrupt the process of DNA replication. Since this replication occurs in an aqueous environment, the changes in the genome should not radically alter the statistical properties of the four nucleotides (their frequencies) and should correspond to the concentrations of these nucleotides in the aqueous solution. The rearrangements in the genome must follow a specific mathematical model. The genome's structure that meets these requirements is described by the Markov chain model, which represents a homogeneous and ergodic process.

The ergodicity of a homogeneous Markov chain means that the probability of each amino acid at any position in the sequence becomes constant fairly quickly (from the beginning of the chain) and equals the frequency of that amino acid, as calculated based on the length of the protein. It has been established that the frequencies of the letters A and T, as well as G and C, are equal for all chromosomes. Thus, the principle of complementarity is not only fulfilled between two opposite strands of DNA but also for each individual strand itself.

Each nucleotide consists of three parts: a monosaccharide, a phosphate residue, and a nitrogenous base. All the nitrogenous bases — A, T/U, G, C — have a flat (planar) configuration. The bases C, G, and T(U) contain an oxygen atom, which adenine (A) does not have. The bases A, T, and G each have 5 carbon atoms in their molecular structure, while C (cytosine) and U (uracil) have only 4 carbon atoms. Thymine differs from uracil in having a methyl group (–CH_3), which is absent in uracil. Thymine (T) is the only one of all the bases that is replaced by uracil (U) in the transition from DNA to RNA. The interchangeability of the letters T and U in DNA and RNA texts is related to a characteristic of the genetic code, as it applies to all living organisms. The other three bases in DNA and RNA chains are the same; namely, A, G, and C each contain an amino group with the structure NH_2, while thymine (T) and uracil (U) do not contain any such amino group at all. Amino groups are a fundamental component for enabling enzymes to recognize the structure of amino acids. Protein synthesis always begins from the 5′-end of the nucleotide chain.

Example from the theory of statistical decision-making:

The original data for making decisions are usually provided by an experimental matrix of conditional probabilities E, values of prior probabilities, and a loss matrix L.

Consider the application of statistical decision theory for identifying three objects: A_1, A_2, A_3, based on the values of a single binary feature (for brevity): x' = 1 and x'' = 0.

Let's consider the given loss matrix L (Table 6.3).

Table 6.3

The Loss Matrix (L)

A	d_1	d_2	d_3
A_1	0	1	1
A_2	C_{12}	0	1
A_3	C_{13}	C_{23}	0

If you choose the correct code, there will be no losses; all other information will indicate the cost of making a wrong decision.

In the matrix L: d_1 corresponds to solution A_1, d_2 corresponds to solution A_2, d_3 corresponds to solution A_3.

The experimental matrix of conditional probabilities is constructed by a statistician based on observations of the distribution of objects. These input data are, in the theory of recognition, called a training sequence. For the selected feature "x", which can assume two values: $x' = 1$ and $x'' = 0$, the experimental matrix of conditional probabilities is presented in Table 6.4. The values are as follows:

- $p(x'/A_1)$ represents the conditional probability (frequency) of the occurrence of a certain effect (the arrival of an amino acid) at the system's input when A_1 is present;
- $p(x''/A_1)$ represents the conditional probability of the occurrence of the input effect x'' when A_1 is present;
- $p(x'/A_2)$ represents the conditional probability of the occurrence of the input effect x' when A_2 is present;
- $p(x''/A_2)$ represents the conditional probability of the occurrence of the input effect x'' when A_2 is present; and so on.

Table 6.4

Experimental matrix (E)

A	$x'=1$	$x''=0$
A_1	$p(x'/A_1) = 0.2$	$p(x''/A_1) = 0.8$
A_2	$p(x'/A_2) = 0.4$	$p(x''/A_2) = 0.6$
A_3	$P(x'/A_3) = 0.8$	$p(x''/A_3) = 0.2$

$p(x'/A_1) + p(x''/A_1) = 1,$

$p(x'/A_2) + p(x''/A_2) = 1,$

$p(x'/A_3) + p(x''/A_3) = 1.$

If the statistics contain no additional information beyond the data from the experimental table, then the prior probabilities $p(A_i)$ are equal to each other, that is:

$p(A_1) = 0.33; p(A_2) = 0.33; p(A_3) = 0.33.$

To make a decision, such as classifying an amino acid into the appropriate category, it is necessary to calculate the expected loss for each input stimulus (i.e., the effect of providing the corresponding feature value to the system for that specific amino acid) and to select the value (for example, the amino acid's structure) that results in minimal losses. The calculation of the loss function is performed using the following formula:

$$\Sigma = \sum_{i=1}^{A}\sum_{k=1}^{q} L[A_i d(x_k)] \cdot p(x_k/A_i)p(A_i) \text{ or}$$

$$\Sigma = \Sigma\Sigma \text{ (LEp)},$$

where A is the number of classes (in our example, A = 3); q is the number of values of the input Boolean parameter [q = 2{0, 1}].

Using this formula, you need to calculate the losses for each possible decision and then select the option that results in the lowest loss. Let's demonstrate how to apply this formula using our example.

Upon receiving the feature value "x'" with a value of 1 as input to the system, we obtain:

$$\Sigma_1 = L(A_1d_1)\cdot p(x'/A_1)\, p(A_1) + L(A_2d_1)\cdot p(x'/A_2)p(A_2) + L(A_3d_1)\cdot p(x'/A_3)p(A_3) =$$

$$= 0 + C_{12}\cdot 0.4\cdot 0.33 + C_{13}\cdot 0.8\cdot 0.33;$$

For example: $C_{12} = C_{13} = 1$ value $\Sigma 1 = 0 + 1\cdot 0.4\cdot 0.33 + 1\cdot 0.8\cdot 0.33 = 0.396$;

$$\Sigma 2 = L(A1d2)\cdot p(x'/A_1)\cdot p(A_1) + L(A2d2)\cdot p(x'/A_2)\, p(A_2) + L(A3d3)\cdot p(x'/A_3)p(A_3) =$$

$$= 1\cdot 0.2 \cdot 0.33 + 0 + 1\cdot 0.8 \cdot 0.33 = 0.33;$$

$$\Sigma 3 = L(A1d3)\cdot p(x'/A_1)p(A_1) + L(A2d3)\cdot p(x'/A_2)p(A_2) + L(A3d3)\cdot p(x'/A_3)p(A_3) =$$

$$= 0.2\cdot 0.33 + 0.4\cdot 0.33 + 0 = 0.198.$$

Since the loss was minimal when $\Sigma 3 = 0.198$ (with $\Sigma 1 < \Sigma 2 < \Sigma 3$), the comparator will point to object (structure) A_3.

When the input value x'' = 0 is submitted to the system, the comparator's inputs will receive signals corresponding to the values:

$\Sigma 1 = 0.264$ (when $C_{12} = C_{13} = 1$); $\Sigma 2 = 0.33$ (when $C_{23} = 1$); $\Sigma 3 = 0.462$.

With such an input, the system will identify the object (structure) A_1.

More details about the role of mathematical methods will be discussed below.

6.2. THE PRINCIPLE OF SYMMETRY AND HARMONY IN THE GENETIC CODE

The natural world around us contains many unknowns and mysteries. Humanity strives to uncover the truth, and this is an inherent objective of human knowledge. Pavel Kotov mentions in the article [45]: "By observing nature, humans have recognized a certain harmony, namely proportionality. It is believed that the components of any system (for example, protons and neutrons in an atom's nucleus) can exist in ratios corresponding to the golden ratio, which is approximately 0.618. Scientists believe that when this ratio is maintained, the system functions optimally. The golden ratio is also observed in nature, even at its most fundamental levels. Take protein molecules, for instance; they are the building blocks of tissues in all living organisms. These molecules differ in mass, which depends on the number of amino acids they contain. It has recently been determined that the most common protein masses are 31, 81.2, 140.6, 231, and 319 thousand units."

The famous Periodic Table of Elements created by D.I. Mendeleev in 1869 not only helped to organize the world of chemical elements but also predicted the discovery of many new elements.

The principle of symmetry is recognized as one of the fundamental foundations of modern natural science. As we have already mentioned, living organisms belong to the category of complex natural systems that cannot be explained by a single, coherent scientific theory in a straightforward manner. Therefore, the principle of symmetry has long been utilized in the field of molecular genetics (for example, in the study of the symmetry of the double helix structure of DNA).

According to E. Chargaff's Rule, we obtain:

1. The number of adenines is equal to the number of thymines, and the number of guanines is equal to the number of cytosines, that is: A = T; G = C. Therefore, A/T = G/C = 1.
2. The number of purines is equal to the number of pyrimidines: A + G = T + C.
3. A + C = G + T, that is, the amount of adenine and cytosine is equal to the amount of guanine and thymine.

Unlike the complementary base pairs A=T and G≡C, the formation of pairs between two purines (A–G) or two pyrimidines (T(U)–C) is sterically hindered. This is because suitable hydrogen bonds cannot be formed, and the geometry of the resulting helix structure is compromised. In the presence of nitrous acid, specific mutations in RNA occur, characterized by the substitution of base A for G and C for U. Consequently, bases A and C are considered "mutable" (i.e., capable of undergoing change), while bases G and U are considered "immutable" (i.e., unable to change).

It has long been known that the creative principle underlying all of nature is harmony. The "Mathematical Theory of Harmony" aims to discover the laws of harmony and possesses a universal character; in other words, it can be applied to all structures in nature at any level of their organization. According to Pythagoras, harmony has a numerical expression, which means it is inherently connected to the concept of number. It is believed that Kepler observed the biological phenomenon of phyllotaxis (the orderly arrangement of leaves on plants) and established a connection between the Fibonacci numbers and the golden ratio, demonstrating that the sequence of ratios of adjacent Fibonacci numbers…

$$\frac{1}{1}, \frac{2}{1}, \frac{3}{2}, \frac{5}{3}, \frac{8}{5}, \frac{13}{8}, \frac{21}{13}, \frac{34}{21}, \ldots.$$

Within certain limits, things strive towards the golden ratio… that is to say…

$$lim_{n\to\infty} \frac{F(n)}{F(n-1)} = \tau = \frac{1+\sqrt{5}}{2} \approx 1.618. \tag{6.7}$$

The Fibonacci numbers are elements of the numerical sequence 1, 1, 2, 3, 5, 8… (the Fibonacci series), where each subsequent number is equal to the sum of the two preceding ones.

In 1990, the French researcher Jacques Peres, who was working as a scientist at IBM at that time, made a discovery in the field of genetic coding. He discovered a mathematical law that governs the self-organization of the nucleotides T, C, A, and G within DNA. He found that sequences of nucleotides in DNA are organized into structures of a higher order, known as "resonances." A resonance represents a particular proportion that determines how DNA is arranged in accordance with the Fibonacci sequence (1, 2, 3, 5, 8, 13, 21, 34, 55, 89, 144, …).

The key idea behind the discovery, which Peres named the "DNA SUPRA-code," is as follows: Take any segment of genetic code consisting of the bases T, C, A, and G; let the length of this segment be a Fibonacci number (for example, 144). If the number of T bases in this DNA segment is 55 (a Fibonacci number), and the total number of A, C, and G bases is also 89 (another Fibonacci number), then this segment of genetic code exhibits a "resonance"—that is, there exists a proportional relationship between these three consecutive Fibonacci numbers (55–89–144). The discovery reveals that every DNA sequence contains multiple such resonant segments. In general, segments of genetic code with a length equal to a Fibonacci number of the form F_n are divided by the "golden ratio" into two parts: one part consisting mainly of T bases (with a specific number of T bases in this segment, as indicated by $F_n - 2$), and the other part consisting of the remaining bases (with a total number of these bases equal to $F_n - 1$). Since 1990, this pattern has been repeatedly verified by numerous biologists. According to the author, the DNA SUPRA-code represents a universal biomathematical principle that indicates the highest level of self-organization of nucleotides within DNA, in accordance with the golden ratio rule [126].

Statistical analysis has confirmed the presence of symmetry in the genomes of bacteria, plants, higher organisms (approximately a hundred genomes in total), including humans. It has been demonstrated that the symmetry of base sequences leads to the symmetry of shorter sequences, as well as of individual bases. Based on the Markov chain model, it was shown that the symmetry of base sequences arises from the symmetry of base pairs. The symmetry of the sequence [127] is explained in the following manner:

1. For the bases that are arranged along a single strand of a DNA chromosome, approximate ratios are applied.

$$n(A) = n(T), \quad n(C) = n(G), \tag{6.8}$$

where n(j) represents the number of bases of type j (j can be {A, C, G, T}) that are counted on a single strand of DNA. It should be noted that, due to the complementary nature of DNA base pairs across the two strands, the numbers of A and T bases, as well as C and G bases, counted on a single strand, do not necessarily correspond to each other.

It follows from equation (6.8) that the numbers for each base, calculated using the first and second threads, are identical.

$$n(A, 1) = n(A, 2), \; n(T, 1) = n(T, 2),$$
$$n(C, 1) = n(C, 2), \; n(G, 1) = n(G, 2). \tag{6.9}$$

The numbers 1 and 2 denote the strand numbers.

This establishes a symmetry with respect to the base sequences on each strand of DNA. Consequently, it follows that the weights of the two strands are identical. Calculations have shown that the following relationships apply to the pairs of bases:

$$n(AC) = n(GT),\ n(AG) = n(CT),\ n(TC) = n(GA),$$

$$n(TG) = n(CA),\ n(AA) = n(TT),\ n(CC) = n(GG). \tag{6.10}$$

The four pairs of AT, TA, CG, and GC are not present here because they would lead to tautologies (repeated or meaningless statements).

1. For bases that are written in the form of formulas, this looks like follows:

$$n(i,j)=n(\bar{j}, \bar{i}), \tag{6.11}$$

where i, j belong to {A, C, G, T}, $\bar{A} = T$, $\bar{C} = G$, $\bar{T} = A$, $\bar{G} = C$.

From equations (6.10) and (6.11), it follows that there is a symmetry with respect to the arrangement of 16 base pairs on each strand of DNA.

$$n(i,j,1) = n(i,j,2), \tag{6.12}$$

where i, j belong to the set {A, C, G, T}.

On the other hand, for the pairs AT, TA, CG, and GC that are not included in (6.10), use the following obvious relationship:

$$n(i) = n(A_i) + n(C_i) + n(G_i) + n(T_i) = n(iA) + n(iC) + n(iG) + n(iT), \tag{6.13}$$

where $i \in \{A, C, G, T\}$, we obtain two connecting constraints:

$$n(CA) + n(GA) + n(TA) = n(AC) + n(AG) + n(AT),$$

$$n(AC) + n(GC) + n(TC) = n(CA) + n(CG) + n(CT).$$

Based on the evaluations of the transition probabilities $p(ij) = n(i,j)/n(i)$ and a pseudorandom number generator, a random base sequence is constructed, matching the length of the human chromosome $\hat{p}(ij) = \frac{n(i,j)}{n(i)}$

The programs for generating random numbers construct a random base sequence that has the same length as the human chromosome. Calculations have shown that the relative difference between successive base triads is less than 1%. Thus, the Markov model confirms the symmetry of these short base sequences.

The authors note that the resulting findings open up numerous possibilities for applying Bayesian procedures to Markov chain models in order to identify genes located on the DNA strands [127].

At the same time, it should be noted that in 1950–1953, E. Chargaff and his colleagues published a groundbreaking study on the chemical structure of nucleic acids. They examined a vast number of different organisms and collected samples from various organs and tissues. The research revealed that the DNA extracted from human cell nuclei consisted of 30% adenine, 20% guanine, 20% cytosine, and 30% thymine. These proportions were observed over lengths of hundreds of base pairs or less; the longer the DNA chain, the smaller (less pronounced) the deviations in these ratios become. For a complete chromosome in higher eukaryotes, an approximate equality can be observed: A ≈ T, G ≈ C, with errors in these ratios amounting to only a few hundredths of a percent.

The analysis of nucleotide sequences revealed that in regions where no significant information is encoded in the DNA, the frequency of combinations of different nucleotides is close to random. In areas with a high number of genes, there are many G and C nucleotides; whereas in regions with a low number of genes, there are many A and T nucleotides. [99].

6.3. SELECTION OF THE MATHEMATICAL MODEL FOR GENOME RESEARCH

Any science, any research—including that in the field of genetics—involves conducting measurements and introducing numerical values. In essence, what is done is to create a mathematical model that allows us to calculate or predict certain outcomes, and then to compare those results with experimental data.

The history of science is full of examples that confirm a "law" formulated by the French mathematician Charles Hermite: "Everything that is mathematically correct will, sooner or later, emerge from its narrow confines and acquire a broader significance."

Thanks to the development of models that study the dynamics of complex living systems, mathematics has had a tremendous impact on biology and medicine.

Based on the knowledge of genetic material and the processes that are occurring, mathematical models are constructed.

A model is a copy or analogy of the process or phenomenon being studied. A mathematical model must meet certain requirements: accuracy, permissible error, universality, cost-effectiveness, and other criteria. In real-world decision-making situations, characteristics are often stochastic and fuzzy (imprecise); therefore, the model's parameters are defined as random variables chosen according to a specified probability distribution law.

When discussing possible approximate approaches to solving the problem of determining the spatial structure of protein molecules, we must consider that all such approaches must be supported by solid theoretical or experimental evidence; without this evidence, the calculation results themselves are of no value. Since millions of chemical reactions occur within cells every minute, gene switching responsible for the production of specific proteins occurs depending on the concentration of certain regulatory proteins. Moreover, living organisms have to deal with inaccuracies in the information provided to them. This may be due to variations in factors such as light intensity and protein concentrations throughout the day.

The analysis of DNA in higher organisms and proteins interests not only geneticists but also mathematicians. To analyze the amino acid sequence of a protein, a probabilistic model of the sequence is constructed.

Any mathematical model describes only a subset of the properties of a particular problem. Therefore, the accuracy of a mathematical model is determined by the degree to which the values of the variables corresponding to real-world objects match the values of those same variables obtained from that mathematical model. All mathematical models of biological phenomena are, by necessity, approximate; no model takes into account all relevant factors. An essential part of any such model is a set of parameters whose values are determined through experimentation, by comparing the predictions of the theory based on that model with actual experimental results.

By comparing the theory with new experimental data, the degree to which the selected model is adequate can be gradually determined. When considering the internal temporal relationships within the developing process, the prediction is merely probabilistic.

By comparing the diversity of both engineering and biological approaches, it is evident that modeling has played a pivotal role in the design and analysis of system properties. For example, the Boeing 777 was fully modeled before being tested in an aerodynamic wind tunnel. Attempts to create mathematical models of individual living cells have been ongoing for a long time; such models would enable experiments to be conducted in order to achieve the desired effect under various conditions affecting the cells.

6.4. THE MARKOV CHAIN AS A MODEL IN GENETICS

A direct generalization of the independent trials scheme is the so-called Markov chain model, which was first systematically studied by the renowned Russian mathematician A.A. Markov.

Markov Andrey Andreevich (1856–1922) was a prominent Russian mathematician and academician of the Imperial St. Petersburg Academy of Sciences. A.A. Markov made significant contributions to the development of number theory, mathematical analysis, probability theory, and mathematical statistics. He also founded the theory of stochastic processes, now known as Markov processes. He proved the Limit Theorem, which was the first rigorously proven result among the so-called ergodic theorems, playing a crucial role in modern physics and engineering.

Ergodicity is a special property of certain dynamic systems, characterized by the fact that, during the evolution of such a system, nearly every point in the system will, with a certain probability, pass close to any other point within the system.

In the theory of Markov chains, a particular issue of interest is the asymptotic behavior of the probabilities P_ij(n). A Markov chain is said to be ergodic if there exists a limit for each value of j.

$lim_{n\to\infty} P_{ij}(n) = P_j$ (which do not depend on i).

For the ergodic chain, the law of large numbers [128] holds true. The law of large numbers is a fundamental principle of probability theory, which states that the combined effect of a large number of random factors, under certain very general conditions, will result in an outcome that is almost entirely independent of chance [46]. This law is expressed through a series of theorems, each proving a particular case of the tendency for the average characteristics of a large number of experiments to converge towards certain constant values. Two well-known theorems in this series are Chebyshev's theorem (1821–1894) and Bernoulli's theorem (1654–1705).

Bernoulli's Theorem about the Law of Large Numbers

Bernoulli's theorem can be stated as follows: "If an unlimited number of experiments are conducted, and in each experiment certain events have the same probability of occurring, then for a sufficiently large number of experiments, it is possible to conclude with a probability arbitrarily close to certainty that the ratio of the number of times the event occurs to the total number of experiments will deviate from the probability of that event by less than any given amount, no matter how small that amount may be."

The name Bernoulli is also associated with the classical definition of probability, according to which the probability of a random event is the ratio of the number of favorable outcomes (m) to the total number of possible outcomes (n):

$$P = \frac{m}{n} \tag{6.14}$$

The idea underlying classical probability was already used by Kepler. He argued that planets "probably" would never return to their original positions at the time of creation, and therefore the end of the world was unlikely to happen.

The definition of classical probability was only introduced by J. Bernoulli in the first chapter of "Ars Conjectandi" ("The Art of Conjecturing"). "Probability, then, is a measure of certainty (probabilitas enim est gradus certitudinis), and it differs from certainty in that it is considered a part of a whole," he stated [129].

In addition to the definition of classical probability that emerged in the second half of the 17th century, statistical (aposteriori) approaches to defining probability were also beginning to be applied in demographic studies. These approaches consider probability as the ratio of the number of occurrences of a random event (denoted by μ) to the total number of trials (denoted by ν).

$$\hat{P} = \frac{\mu}{\nu} \tag{6.15}$$

J. Bernoulli wondered whether it might happen that, as the number of observations increases, our knowledge would still be limited to a certain degree of certainty.

Let p and $\hat{p}$ denote the classical probability and the statistical probability of some event, respectively. Then Bernoulli's doubt can be expressed as follows: Do there not exist such positive numbers β and δ that the limit of the probability as $\nu \to \infty$ is...

$$P(|p - \hat{p}| < \beta) \text{ less than or equal to } 1 - \delta?$$

Using his theorem, Bernoulli answers this question in the negative: no, such β and δ do not exist. In doing so, he demonstrated the equivalence between statistical probability and classical probability, and, within the framework of the probabilistic theory of knowledge, established a correspondence between the inductive and deductive methods. He showed how statistical observations gradually lead to a well-founded belief (or, in other words, to moral certainty).

The Law of Large Numbers

According to the modern common definition, a sequence of random variables...

$$\xi_1, \xi_2, \ldots\xi_n, \ldots$$

It follows the "law of large numbers" if there exists such a numerical sequence: $a_1, a_2, \ldots, a_n, \ldots$.

$$lim_{n\to\infty} P\{ \left| \frac{1}{n}\sum_{k=1}^{n} \xi_k - a_n \right| < \varepsilon \} = 1. \tag{6.16}$$

The fairness of the "law of large numbers" for the Markov chain process is one of the fundamental principles of probability theory.

Markov chains have seen further development in science and are now widely used in physics for the study of gas diffusion. In engineering, Markov chains are used to describe certain information transmission processes, a range of technological procedures, as well as systems for monitoring operational performance and detecting faults in complex technical systems. In sociology, they help analyze changes in the social or occupational structure of populations and issues related to population migration. In biology, Markov chains are utilized to study the development patterns of various animal and plant species.

Example from cellular genetics: [130].

A Markov chain with the set of states $E_0, \dots, E_j, \dots, E_n$ and the corresponding transition probabilities...

$$P_{jk} = \binom{2j}{k}\binom{2N-2j}{N-k}\binom{2N}{N}^{-1} \qquad (6.17)$$

This phenomenon appears in a biological problem that can be simplified as follows: Each cell of a certain organism contains N particles, of which some belong to type A and others to type B. A cell is said to be in the state E_j if it contains exactly j particles of type A. Daughter cells are formed through cell division, but before division, each particle replicates (i.e., its number doubles). The daughter cell inherits N particles randomly selected from the 2j particles of type A present in the parent cell, as well as 2N – 2j particles of type B. The probability that a daughter cell will be in the state E_k is then given by the hypergeometric distribution (6.17).

The development of communities of living organisms can naturally be viewed from a theoretical perspective as a Markov process, assuming that we possess complete genetic information for the initial moment in time. In the article "On the Issue of Modeling Evolutionary Processes," O.S. Kulagin and A.A. Liapunov noted: "When considering a natural process that evolves over time, the question arises as to what its internal temporal relationships are—that is, what information is needed in order to predict its future course. It is not assumed that this process unfolds in a deterministic manner; therefore, the prediction is probabilistic. A process is considered Markovian if comprehensive data for making a prediction can be obtained by describing the system's instantaneous state; additional information about previous states does not contribute to the predictive outcome. It is important to note that the same process may be Markovian under one set of descriptive criteria but not another (often indicating that the second set of criteria is flawed). For example, the problem in celestial mechanics is Markovian if the positions, velocities, and masses of the bodies are known; it is not Markovian if only the positions and masses are given. When discussing the development of a community of living organisms under specific conditions, this process is not Markovian when described purely from a phenotypic perspective, as information about recessive genes is lacking. However, when a genotypic description is used, the process becomes Markovian. Certain insights into changes in external conditions can affect the probabilities of the system transitioning from one state to another; yet additional information about past external conditions or genotypes has no impact on the probabilities of future transitions. Thus, from a theoretical standpoint, the development of communities of living organisms can naturally be regarded as a Markov process, assuming we have complete genetic data for the beginning of time." [131]

A.A. Markov understood a sequence of random numbers, where the probabilities of these numbers occurring are interrelated—more precisely, the probability of each subsequent number being determined by the previous one. Such mathematical sequences were later named "Markov chains" in science. For many years, Markov chains did not find significant practical applications. However, they have recently gained popularity again and, along with other types of mathematical models, occupy a prominent place in modern research. Concepts such as simple and complex, homogeneous, and non-homogeneous Markov chains have been defined. Moreover, Markov chains have given rise to a new major branch of probability theory: the theory of stochastic processes. If it is assumed that the transition probabilities remain constant throughout each experiment (operation), then the process can be described by a simple, homogeneous Markov chain. For homogeneous Markov chains, the probability of the system transitioning from state E_i to state E_j in one step depends solely on the starting and ending states; in other words, the transition probability matrix is independent of the step number.

The simplest model of a chain can be represented as freely connected nodes. Such a chain can be described as a sequence of vectors, where each vector indicates the direction from one previous node to the next. A node may contain multiple monomers of the polymer chain. The key point is that each node can rotate freely in its connection relative to its neighboring nodes in the chain.

Let's consider the interpretation of Markov chains in the context of molecular genetics. First, let's take a molecule that has been predetermined in advance. We just need to agree on which specific molecule we are dealing with. It is believed that molecules can have either linear or branched structures; moreover, the branching patterns can potentially involve an infinite number of combinations. Let's start with the simplest linear form. This type of structure corresponds to the primary structure of the DNA molecule. A linear macromolecule can be represented as a long chain consisting of various segments (or "links") of nucleotide sequences.

Let's represent the chain as follows:

T-C-C-T-T-T-C-A-A-G-A-C-C-T-A-G-A-A-G-G-T-C-C-A

In order to understand the structure of a molecule, it is necessary to determine the number of various types of bonds and their arrangement order. This can be done by moving along the molecule and counting specific sections. A probabilistic relationship between these sections is characteristic of macromolecules. When macromolecules are formed from monomers (the original building blocks), each subsequent bond has a certain tendency to either form or not form with the previous bond. This phenomenon is known as the "end effect." Therefore, the probability of encountering the next bond will depend on which section of the molecule we are currently at.

Let's consider the chain we are examining, which consists of nucleotide sequences. Since the chain is made up of four types of nucleotides—A, T, G, and C—there will be a total of 16 different such segments: AA, AG, AT, AC, GA, GG, GT, GC, TA, TG, TT, TC, CA, CG, CT, CC. If we already know the probabilities of transitioning from one segment to another (for example, from AA to AG, etc.), then a complete probabilistic description of a single transition step in the chain can be represented using a transition matrix (Table 6.5). The P matrix has the following properties:

a) $0 \leq P_{ij} \leq 1$;

b) $\sum_{j=1}^{k} P_{ij} = 1 \qquad (i = 1, 2, \dots, k),$

In other words, the sum of the elements in each row of the transition matrix equals one.

Square matrices that satisfy conditions a) and b) are called stochastic (random) matrices. The vector $a = (a_1, a_2, \dots, a_k)$ is defined, where $a_i = P(E_i)$ represents the probability of the state E_i ($i = 1, 2, \dots, k$) occurring in the initial experiment; this vector is known as the vector of initial probabilities. The properties of homogeneous Markov chains are completely determined by the initial probability vector and the transition probability matrix.

We need to make a clarification in our reasoning. For example, P(AA) should not be considered as a probability of transitioning to the same area A again, if we are still in that area; rather, it represents a new area A, similar to the one we were in before. The states are irreversible because from any of these areas, we can move to any other area and continue further, but we cannot return back to the previous state.

Then, the beginning of our path should also be represented by the distribution of probabilities for the four letters: P (A), P (G), P (T), P (C). Using a transition matrix, we can determine how many times, on average, we move to a particular state (node) before reaching the end of the chain.

If we know when and how many times a particular nucleotide will occur along the path of this molecule, and we sum these values for all molecules in the polymer, then the problem is solved—i.e., the composition and structure of the polymer can be determined. The average number of occurrences of each of the four states (A, G, T, C) will not depend on the initial probabilities. Such properties are characteristic of a regular Markov chain. Recall that if a stochastic chain does not exhibit cyclic behavior after any number of steps, it is called a regular chain. This means that in a regular chain, transitions can occur at any step, and the transition probabilities in the transition matrix must be non-zero at any step along the chain. Furthermore, it turns out that combinations of states may not repeat after a certain number of steps. Another property of regular chains is that the matrix assumes a stationary form after a relatively small number of steps (Table 6.5); in this case, the sum of the probabilities across all rows equals one, and the initial probabilities of the letters also add up to one.

Table 6.5

	A	**G**	**T**	**C**
A	P(AA)	P(AG)	P(AT)	P(AC)
G	P(GA)	P(GG)	P(GT)	P(GC)
T	P(TA)	P(TG)	P(TT)	P(TC)
C	P(CA)	P(CG)	P(CT)	P(CC)

For the stationary case, we can only determine the average number of each type of dinucleotide segment (e.g., AA, AG, AT, AC, GA, GG, GT, GC, TA,

TG, TT, TC, CA, CG, CT, CC) in the molecule; in other words, we can only ascertain the composition of the polymer. However, its structure—i.e., the specific order of the nucleotide segments—is still unknown. To address this, we need to divide the chain into larger segments. It is more convenient to divide the chain into segments of three nucleotides each, since amino acids are composed of triplets (three nucleotides). Thus, our chain will consist of sequences of amino acids, which in turn correspond to sequences of three nucleotides each. Alternatively, we could divide the entire chain into even larger segments, for example, of ten nucleotides each. A segment of this type might look like "TCCTTTCAAG". Knowing the number and type of such segments in the chain would allow us to determine the complete composition and structure of the macromolecule as well as the polymer as a whole.

The estimates for the transition probabilities of the Markov chain are presented as follows:

$$\hat{P}(ij) = \frac{n(ij)}{n(i)}, \tag{6.18}$$

The symbol "P" with a "hat" is used to represent probability estimates.

n(ij) – the number of amino acid pairs (ij) in the protein.

n(i) – the number of amino acids.

The ergodicity of a homogeneous Markov chain implies that the probability of each amino acid in the sequence, starting from the beginning of the chain, becomes constant and equal to the frequency of that amino acid.

In different proteins, the frequencies of amino acids (and consequently, the estimates of transition probabilities) may vary.

6.4.1. Bayesian Recognition Procedure for Markov Chains

Bayes' formula is the foundation of the inductive approach to recognition. The inductive method, as a way of reasoning from the particular to the general, represents an alternative to the axiomatic-deductive approach. While the deductive approach leads us to a true conclusion, we must rely on concepts of probability in order to justify inductive conclusions.

The recognition process is based on comparing the features of the object under study with the features of other known objects, in order to determine the most likely correspondence between them. These recognition methods are applied in technical diagnostics, criminal investigations, expert systems, production process management, and various other fields.

This research direction was initiated by the work of the American neurophysiologist F. Rosenblatt, who in 1957 developed a model of brain perception called the "Perceptron" [132].

In the USSR, significant contributions to the development of theory and practice in the field of recognition were made by scientists such as V.M. Glushkov, N.P. Buslenko, V.S. Mikhalevich, A.G. Ivakhnenko, and Yu.I. Zhuravlev.

Currently, the interest in researching recognition methods has somewhat diminished. This is because many of the existing recognition methods have been well-studied and are widely used in practice. However, their adoption is progressing more slowly than expected.

The issue related to estimating the errors in recognition procedures has not been sufficiently explored. A key aspect of the theory of error estimation is the structure of the samples and their probability distributions.

It is known that the main chain of a protein, like the polynucleotide chain, possesses connectivity. The peptide bond (—CO—NH—) is a chemical bond that connects the amino group of one amino acid to the carboxyl group of another in peptides and proteins.

A graph, which is an analog of a polypeptide chain, looks as follows:

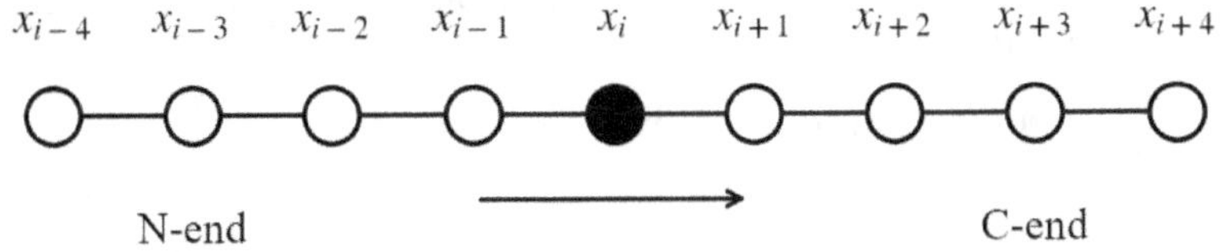

The vertices of the graph represent amino acids (represented by circles), and the edges (represented by solid lines) represent peptide bonds. The direction of the chain (indicated by the arrow) goes from the N-terminus to the C-terminus, and it corresponds to the sequence in which amino acids are attached during protein synthesis. If the amino acid being analyzed at the current moment is denoted by the letter x_i (the circle is colored), then all amino acids preceding it will be denoted as x_{i-1}, x_{i-2}, x_{i-3}, and so on. Similarly, all amino acids following x_i will be denoted as x_{i+1}, x_{i+2}, x_{i+3}, and so on. Atoms can form bonds due to their electronegativity, as indicated by the symbol [46].

The probabilistic relationship between the various parts mentioned above is characteristic of macromolecules. When macromolecules are formed from monomers, each subsequent unit tends to either readily or less readily attach to the previous one; this phenomenon is known as the end effect. Therefore, the choice of the next unit will depend on the level at which the preceding unit currently resides.

Let's consider a simple "monochain" element:

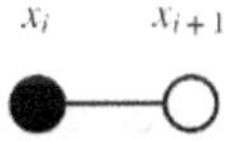

Let's denote x_i as letter A, and x_{i+1} as letter B. The monomer chain will represent the connection:

$$A \xrightarrow{k} B,$$

where k is the proportionality coefficient (the rate constant).

Let's determine the probability of the interaction between amino acid A and some other amino acid B. Clearly, the rate at which amino acid B binds to amino acid A is determined by the likelihood of the pair being in a "reactive" state—that is, a state in which amino acid B is able to bind, and amino acid A is also able to accept the binding.

Let's introduce the following concepts:

p(B,t) – the probability that event B occurs, namely that an amino acid is in a state suitable for binding at time t.

p(A, t) – the probability of event A, that is, the probability that at time t amino acid A is in a state capable of binding.

p(BA, t) represents the probability that the first two events occur simultaneously (in probabilistic terms, this is called the "product" of the two events), or the probability that the amino acid pair in question is in a "reactive" state at time t. According to the theorem of probability multiplication, the probability of the product of two events, p(BA, t), is equal to the product of the probability of one of the events multiplied by the conditional probability of the other event, assuming that the first event has already occurred (denoted as [114]). In other words, p(BA, t) is the probability of finding the BA amino acid pair in a reactive state at time t.

$$p(BA, t) = p(B, t)\, p(A|B, t)$$

or (6.19)

$$p(BA, t) = p(A, t)\, p(B|A, t),$$

where p(A|B,t) is the conditional probability of finding amino acid A in a state ready to bind, given that amino acid B is also in a state ready to bind; similarly, p(B|A,t) is the conditional probability of finding amino acid B in a state capable of binding, given that amino acid A is in a state ready to bind.

Obviously, the probabilities p(B,t), p(A|B,t), and p(B|A,t) considered in this context are functions of time: the corresponding amino acids remain in their initial, unmodified states until the process of attaching B to A occurs. The Bayes' theorem can be easily derived from the theory of probability multiplication. It will look as follows:

$$p(B|A,t) = \frac{p(B,t)\, p(A|B,t)}{p(A,t)}. \quad (6.20)$$

The rate at which the amino acid B binds to A is proportional to the likelihood that the particular pair of amino acids is in a reactive state (i.e., capable of participating in the chemical reaction).

In a real-world system, there are certain amino acids that are in states that allow them to bind to amino acid A. To transition to such a system, it is necessary to sum up and average all the identical amino acids that are part of the environment surrounding the terminal amino acid A.

The above "biomolecular" description of the amino acid addition process is approximate, as it is based on the assumption that any j-th amino acid of type B (out of a total of n) can interact with each of the m amino acids of type A. In other words, what is intended is to sum up all the possible "overlaps" between the elementary chains that make up the main Markov chain. The error in such a description is obviously smaller the more "overlaps" there are between the actual process and the described procedure. Since amino acids differ in their chemical properties and can occupy any position within a polypeptide chain, there are 20^n possible variations for a protein consisting of n amino acids.

In work [133], a procedure for predicting the secondary structure of proteins was investigated, using Bayesian recognition methods based on Markov chains.

It is known that the properties of Markov chains are completely determined by the initial distribution and the transition probabilities.

A Markov chain can be represented in the following manner:

$$x_0 \rightarrow x_1 \rightarrow \ldots \rightarrow x_n.$$

This Markov chain is defined by the initial distribution x_0 and the transition probability matrix. If the transition probabilities $P_{ij}(n)$ do not depend on the step number, then the Markov chain is called homogeneous, which means that $P_{ij}(n) = P_{ij}$ for all values of n. Otherwise, the chain is non-homogeneous.

For a homogeneous Markov chain, the probability of a sequence $x_0, x_1, \ldots, x_n$ is determined by the following relationship:

The value of $x_0, x_1, \ldots, x_n$ is determined by the following relationship:

$$P(x_0, x_1, x_2, \ldots, x_{n-1}, x_n) = P(x_0)\, P(x_0, x_1)\, P(x_1, x_2) \ldots P(x_{n-1}, x_n), \quad (6.21)$$

$P(x_0)$ represents the initial state probability, $P(\chi_i\text{-}1, x_i)$ represents the transition probabilities, where $i = 1, 2, \ldots, n$.

The procedure for predicting the secondary structure of a single amino acid based on Bayes' formula can be described as follows:

$$P(f \mid x1, \ldots, xn) = \frac{P(x1,\ldots,xn \mid f) \cdot P(f)}{P(x1,\ldots,xn)}. \quad (6.22)$$

Here, "f" represents the state of the amino acid. The type of its secondary structure is determined by the surrounding amino acids $x_1, x_2, \ldots, x_n$ located to the left and right of the amino acid under consideration. The probabilities of the sequences $x_1, x_2, \ldots, x_n$ were calculated using Markov chains models for non-stationary processes, according to the following formula:

$$\hat{P}(ij) = \frac{n(i,j)}{n(i)}. \quad (6.23)$$

It is known that the relationship where n(i,j) represents the number of amino acid pairs (ij), and n(i) represents the number of occurrences of the amino acid i in the sequence, serves as an estimate of the transition probabilities for homogeneous Markov chains. The value [133] indicates that for longer chains, these estimates converge to the actual transition probabilities. Consequently, the transition probabilities $P(x_{i-1}, x_i)$ are replaced by their estimated values. The Bayesian recognition procedures for Markov chains of a certain order are thus based on these estimated transition probabilities.

The problem of classification can be viewed as a problem of "statistical decision functions." A researcher may have several hypotheses, each corresponding to a different probability distribution for the observed data. Among all these possible hypotheses, the researcher must first choose one and reject the others. If there are only two hypotheses in the general population (the set of all possible cases), then the task of testing them is simplified: only one hypothesis is tested, corresponding to a specific probability distribution, while another competing hypothesis is considered. In some cases, the categories can be predetermined if the probability distributions of the measurements are fully known; in other cases, the type of each distribution may be known, but the parameters of these distributions are unknown and must be estimated based on a sample from the general population.

6.4.2. Basic principles of the Bayesian recognition procedure

In recent years, the most widely used models have been those that employ artificial intelligence systems for recognition, as well as models based on logical calculations.

Each type of artificial intelligence system has its own characteristics, such as in terms of learning capabilities, generalization abilities, and the ability to draw conclusions. These features make a particular system more suitable for solving certain types of problems and less suitable for others. Neural networks are good at image recognition tasks, but it is quite difficult to understand how they actually perform this recognition. Moreover, it is not possible to feed any prior knowledge (expert information) into a neural network in order to speed up the learning process.

On the other hand, systems based on logical computation are used to explain the conclusions drawn from those computations; however, they cannot automatically acquire knowledge to be used within the conclusion-making processes.

The Bayesian procedure eliminates these shortcomings and optimally solves recognition tasks.

Using Bayes' formula, we can recalculate the probability more accurately by taking into account both the information that was already known and the data from the new observations.

The Bayesian procedure can be applied to both independent features and linearly ordered (dependent) features. When using dependent features, the impact coefficients of these features must be taken into account. It is particularly noteworthy that this recognition procedure can be widely utilized in various fields of science and technology.

To solve recognition problems, the class of objects under study can be represented as a model in the form of a set of points in a feature space. The coordinates of these points correspond to certain properties of the objects. Each coordinate can assume one of two values: 1 or 0. If a particular property is present in an object, the corresponding coordinate takes the value 1; otherwise, it takes the value 0. Such a space is called a Boolean space.

Of course, these models are limited; real objects always possess a multitude of other properties that are not taken into account in such simplifications. However, any model is designed to address a specific set of problems, and from that perspective, many of the properties of real objects turn out to be insignificant, allowing us to exclude them from further consideration.

The Bayesian recognition procedure utilizes Boolean functions, which enables the use of all object classes.

The study of objects is based on measurements obtained using various measuring devices. For each observed object, there is a description of the object in the form of a vector:

$$x_1, \ldots, x_j, \ldots x_n, f.$$

Here, $x_1, \ldots, x_j, \ldots, x_n$ represent the measurements of an object, while f denotes the value of a characteristic of interest to the researcher; this characteristic reflects the state of the object. The variables $j = 1, \ldots, n$ represent the values of various parameters. Often, the values x_j and f are referred to as features (or characteristics).

If there are m such descriptions, then they all together constitute the training set. Here, m_1 represents the size of one class; m_0 represents the size of another class; and $m = m_0 + m_1$ represents the total size of the training set. The structure of the training set is shown in Tables 6.6 and 6.7.

Table 6.6

This is a class for objects from the training dataset that possess a certain property of those objects.

Object number	Measurements					
	x_1	x_2	x_3	...	x_n	f
1						1
2						
...						
m_1						

Table 6.7

The class of objects from the training set for which a certain property is not present in those objects.

Object number	Measurements					
	x_1	x_2	x_3	...	x_n	f
1						0
2						
...						
m_0						

The goal is to construct an inductive reasoning procedure that, based on the measurements $x_1, x_2, \ldots, x_n$ as well as the training dataset, determines the value of the characteristic f.

When studying objects based on a finite number of measurements, the vector of measurements $x = (x_1, \ldots, x_j, \ldots, x_n)$ may correspond to multiple values of the characteristic f. In other words, an object x can simultaneously belong to both classes of objects.

In order to distinguish between the description data, it is necessary to assign a probability distribution to each of the object descriptions.

For reference, if $P(A) > 0$, then the fraction:

$$P(B|A) = \frac{P(AB)}{P(A)} \tag{6.24}$$

This is referred to as the conditional probability of event B occurring given condition A.

It follows from the theorem of probability multiplication that...

$$P(AB) = P(A)\,P(B|A) = P(B)\,P(A|B), \tag{6.25}$$

In other words, the probability that two events occur simultaneously is equal to the product of the probability of one of the events occurring multiplied by the conditional probability of the other event occurring, given that the first event has already taken place.

From this theorem, we obtain:

$$P(A|B) = \frac{P(A)P(B|A)}{P(B)}. \quad (6.26)$$

This is a well-known Bayes' formula that allows us to evaluate different events A based on their relationship with a fixed event B.

Probability theory, like any mathematical science, is an axiomatic theory. The statement (6.26) is actually an axiom regarding the multiplication of probabilities. Therefore, it can be said that most of the results in probability theory are logical consequences of Bayes' formula and the principle of independence of events.

In the cases being considered, event B occurs when the measurements of the object take on specific values: $x_1, \ldots, x_j, \ldots, x_n$. Event A is defined by the characteristic f taking the value f = 1, while event $\bar{A}$ is defined by f taking the value f = 0.

In order to obtain meaningful results, certain assumptions need to be made regarding the probable probability distributions. The probability distributions must satisfy the following condition:

$$P(x1, \ldots, xn \mid f = i) = \prod_{j=1}^{n} P(xj \mid f = i), \; i = 0, 1. \quad (6.27)$$

This fact indicates that the features x_j are independent for each class of objects; here, $P(x_j|f = i)$ represents the conditional probability. The Bayes' formula then links these conditional probabilities together to establish a relationship of dependence.

$$P(f = i|x1, \ldots, xn) = \frac{P(f=i)\prod_{j=1}^{n} P(xj|f=i)}{P(x1, \ldots, xn)}. \quad (6.28)$$

In the Boolean case, the precise evaluation of the recognition procedure's error has a very simple format: [116].

$$a\sqrt{\frac{(n+1)}{m}}, \quad (6.29)$$

where a > 0 is a constant.

"n" is the number of measurements of the object.

$m = \min(m_o, m_1)$ – this represents the size of the smallest class in the training dataset.

The Bayes' formula allows us to determine the posterior probability of the existence (or non-existence) of certain entities (e.g., $A_1, A_2, \ldots$, Ar) given the presence of independent features $x_1, \ldots, x_j, \ldots, x_n$.

By calculating the product of the conditional probabilities and the prior probabilities, taking into account all classes of the objects being studied, the system generates prototypes and then identifies the missing element (if any) for which the probability will be the highest.

The Bayes formula allows us to evaluate various events $A_1, A_2, \ldots$, Ar based on their relationship with the fixed event B.

In the Bayesian approach, the distinction between a parameter and its estimate is reflected in the utility function. The loss function, by its meaning, characterizes the losses that arise from replacing the true parameter with its estimate. The estimation rule is chosen such that it minimizes the mathematical expectation of this loss function. From a Bayesian perspective, the parameter itself is considered random, whereas its estimate is not. The mathematical advantage of the Bayesian approach lies in the fact that it does not require consideration of properties such as efficiency, unbiasedness, or consistency. These properties (especially the first two) are often contradictory for many classical estimation methods; moreover, in general, it is unknown which type of estimate (unbiased or biased but more efficient) should be preferred. Instead of relying on these classical properties, Bayesian theory imposes a single overarching requirement: the estimate should be optimal in the sense of minimizing expected loss. By examining the Bayesian approach from the perspective of the classical property of consistency, one can conclude that Bayesian estimation theory effectively addresses the problem of working with small sample sizes (Reference [51]).

The Bayesian approach is based on the use of a formula. In its simplest form, this formula looks like follows:

$$P(A_i|B) = \frac{P(A_i)P(B|A_i)}{\sum_{i=1}^{n} P(A_i)P(B|A_i)}, \qquad (6.30)$$

Where: $P(A_i)$ represents the prior probability, and $P(B|A_i)$ represents the conditional probability.

If the probability distributions for each class are known, the desired algorithm can be expressed in an explicit, analytical form, and this algorithm will be optimal since it will have the minimum possible probability of error. However, in practice, the probability distributions for the classes are not always known; instead, they must be estimated based on training datasets. In this case, the Bayesian approach cannot necessarily be considered optimal, as the estimated distributions will contain some degree of uncertainty (i.e., errors).

According to the Bayesian approach, using formulas like (6.30), the posterior probabilities of events (A_i) are determined. Now, based on these posterior probabilities, we need to identify which event from system A actually occurs. In the simplest case, the event with the highest posterior probability, namely A_i, can be assumed to be the true event. This evaluation function is called the "maximum a posterior probability estimate." For each A_i that is selected as the true value, a measure of the risk associated with that choice or the corresponding losses is determined. The choice of the event deemed to have actually occurred is made based on the minimum loss. Thus, the Bayesian decision is one that minimizes the average risk value [135].

The task of classification involves selecting a decision rule that minimizes a certain function (in our case, this is the average risk). Recall that the function I(x, u), used as a quality criterion for the chosen dependency, is a rule that assigns a real number to each pair of functions x(t) and u(t). In many optimal control problems, this function is represented in the form of a specific integral.

$$I(x,u) = \int_{t_0}^{T} F(x_1, x_2, \dots, x_n, u_1, u_2, \dots, u_m)\, dt \qquad (6.31)$$

where $F(x_1, x_2.., x_n, u_1, u_2.., u_m)$ is the given function.

Minimizing a function based on empirical data is a task in mathematical statistics. In [136], this is referred to as the problem of minimizing the average risk using empirical data.

The task of training an image recognition model was formulated in the late 1950s. The essential idea behind this task is for the teacher (the classifier) to observe the various situations that arise and determine to which of the k classes each situation belongs. The goal is to create an algorithm that, after observing the data, can perform classification in a similar manner to the teacher. In [136], such a dependency relationship is described using formal language, and its expression is quite simple. In a certain environment characterized by a probability distribution density P(x), situations x occur randomly and independently. The teacher assigns these situations to one of the k classes. For simplicity, the author chooses k = 2 in this example. The classification is then performed using a conditional probability distribution function P(ω|x), where ω = {0;1} (ω = 0 indicates that the teacher assigned situation x to the first class, and ω = 1 indicates that it was assigned to the second class). All the functions F(x,α) belonging to a class are "characteristic" functions, meaning they can only assume two values: zero or one. The goal is to find a set of (ℓ pairs of (x_i, ω_i);…; (xℓ, ωℓ)) – where x_i represents a situation and ω_i represents the teacher's response to that situation – and to select a characteristic function F(x,α) from these functions such that the probability of a misclassification (i.e., a classification different from the teacher's decision) is minimized. In other words, the objective is to minimize the value of the corresponding functional measure.

$$I(\alpha) = \sum_{\omega=0,1} \int (\omega - F(x,\alpha))^2 P(\omega|x)P(x)\, dx. \quad (6.32)$$

It is recommended to write the functionality of I(α) in the following manner:

$$I(\alpha) = \int_{x,\omega} (\omega - F(x,\alpha))^2 P(x,\omega)\, dx\, d\omega, \quad (6.33)$$

The function P(x, ω) = P(ω | x)P(x) is proposed to be called the joint density of the pairs (x, ω), which are defined in the space Xω.

Thus, the task of training for image recognition was reduced to the problem of minimizing the average risk based on empirical data. A special feature of this approach is that the class of functions Q(z,α) does not have the same degree of flexibility as in the general formulation; instead, it is subject to certain constraints.

1) The z-vector consists of n + 1 coordinates: the coordinate ω, which can only take two values (zero and one), and the n coordinates $x_1, \ldots, x_n$, which together form the x-vector.
2) The class of functions is $Q(z, \alpha) = (\omega - F(x,\alpha))^2, \ldots)$

Here, F(x,α) also only takes two values: zero and one.

In the task of image recognition training, the value of the loss function is either zero or one. This characteristic of the risk-minimization problem defines the specific nature of image recognition training. The evaluation of recognition methods should take into account the class sizes, rather than the total length of the dataset. Methods for minimizing empirical risk are inherently inefficient/impractical in their approach: the optimal parameters for the decision rules are determined by minimizing empirical risk on a random subset of training data. Clearly, for a sequence of new objects that are not present in the training data, the previously found parameters may no longer be optimal, and they need to be recalculated. This leads to the issue of dividing the dataset into a training set and a validation (control) set. Methods that involve labor-intensive procedures for tuning optimal parameters on individual training datasets are unsuitable for use in living nature (biological systems), as well as for recognizing objects in rapidly changing environments. However, it has been shown that by averaging over all possible training datasets, one can obtain deterministic estimates of the error rate of the Bayesian recognition procedure, depending on class sizes, the number of features, and their values. These estimates will differ significantly from those obtained by methods that minimize empirical risk.

The schematic diagram of the recognition system that implements the Bayes' formula is shown in Figure A.3.1 of Appendix 3.

6.4.3. System construction and training

The design of a recognition system is a rather complex process that requires the construction of a mathematical model for the system being designed.

To prepare the system for training, it is necessary to convert continuous features into binary values ("1" and "0") and to construct a functional model of the object. With binary features, the size of the system as well as the time required for proper recognition are significantly reduced. This can often be achieved without significant information loss, provided that the threshold value for the continuous variable is chosen wisely. Sometimes, it is necessary to select two threshold values…

if $n < x < v$, we set $x = 0$;
if $x < n$ or $x > v$, we set $x = 1$;
where n and v are the lower and upper threshold values, respectively.

The variety of parameters that need to be controlled, along with the requirement to convert them into a uniform format (in this case, binary) for processing, necessitate the transformation of the signals.

In radio-electronic systems, parameters such as phase, frequency, time intervals, etc. need to be converted into a common format, for example voltage. The voltage can then be easily converted into a binary form by selecting a threshold value for the controlled quantity (as mentioned above).

So, going forward, when we talk about sets, we will usually be referring to sets of binary words, because in the end, a computer can only process elements from such sets.

First, we need to divide a certain set of V elements into N non-overlapping subsets, namely $K_1, K_2, K_3, \ldots, K_n$. We will refer to these subsets as "classes". By "samples," we mean the explicitly defined, non-overlapping subsets of the set V_i that consist of representatives from the classes $K_1, K_2, K_3, \ldots, K_n$. The task is to find algorithms that can determine for any element U in the space V to which class it belongs, and to ensure that it belongs to exactly one class. At the same time, it must be ensured that all elements of the samples remain within their respective classes. By choosing any arbitrary way of dividing a set into classes as a basis, you can construct an algorithm that solves the given problem. The samples of class elements are selected by specialists and experts who have expertise in the subject area of the research. Almost always, there is an intuitive or expert-based way to determine the degree of similarity between one real object and another.

The characteristics of the functional model must be taken into account when preparing the system for training, which involves implementing Bayes' formula for recognition. During the training phase, the recognition system works together with a "teacher." The teacher's role is to present the system with training examples from all the identified classes and indicate to which classes these examples belong. Afterwards, the teacher tests the recognition system by providing it with validation examples. If the system correctly identifies the validation examples, it indicates that it has successfully learned the relevant patterns. If necessary, the teacher increases the size of the training dataset until the average number of errors falls below a desired level. In this way, the system develops the ability to detect even very subtle differences between the recognized objects.

6.4.4. Reading a protein sequence that consists of a chain of nucleotides or amino acids

In order to identify the structure of a chain, it is first separated from the other parts and subjected to enzymes that break the chain into strictly defined segments. Protein enzymes "cut" the amino acid chain at specific locations. Chemistry is capable of distinguishing between different short segments of the amino acid chain with great accuracy; therefore, these segments are subjected to chemical analysis. This process determines the sequence of the amino acids. Next, the chain is reassembled from the identified segments. Then the same protein molecules are taken and exposed to other enzymes, which cause them to break down into further segments whose structures can be studied. By examining the interactions between these individual segments, it is possible to determine the overall sequence of the amino acids in the entire chain.

We will obtain short peptide segments that contain several amino acids. How a protein can be "read" will be demonstrated using an example of a textual sequence in the Russian language. When decoding written text, it is very important to separate vowel letters from consonant letters; combinatorics comes in handy here.

Here's a simple example: Suppose we have some text and we don't know its exact beginning. To "read" the text, we divide it into segments similar to amino acids; each segment consists of three letters.

For example, let's divide the text into groups of segments. The first group consists of the following segments: 1) "sto", 2) "oli", 3) "lic", 4) "ica". The letter "o" appears at the end of segment 1) and at the beginning of segment 2). Therefore, these segments are connected as follows: 1)----2).

However, for the combination of letters: the "l" from the second segment appears in the third segment, followed by another "l". This creates a new sequence: 1)---2)---3). Then we see that at the beginning of the fourth segment, there is a combination of the letters "ica". Another sequence is added to this chain. We get: 1)----2)----3)----4). As a result, the word "Ctolica" (Capital City) is formed.

Similarly, a protein sequence is read as a string of amino acids. The protein chain is then subjected to combinatorial analysis. Accurate combinatorial analysis requires the use of powerful computational techniques.

The method for reading protein sequences was proposed by Fred Sanger in 1977 (British Medical Research Council). For sequencing (determining the sequence of nucleotides), Sanger used radioactive markers and labels corresponding to the number of nucleotides. These labels identified the locations in the DNA sequence where the respective nucleotides were present. The process was later automated by geneticist L. Hood in 1986. Wherever Sanger had used radioactive markers, Hood developed a new method that employed fluorescent dyes of different colors, one for each DNA nucleotide. In Hood's automated device, a laser beam caused the dyes to emit light of various colors; this light was detected and analyzed by a computer. By 1999, the fully automated instrument could sequence up to 150 million base pairs per year [97].

A classic example from linguistics is the Russian phrase equivalent to "Execute cannot pardon" (Убить нельзя помиловать). Depending on where the comma is placed, it can mean either "Execute, cannot pardon" or "Cannot execute, pardon" — without punctuation it is unclear whether the sentence orders an execution or a pardon. The same principle applies in genetics.

In a protein chain, a segment of DNA can be read with a phase shift of either +1 or –1 nucleotides. There are three reading frames within a single strand of DNA. One of these reading frames is considered the correct one; in this frame, the sequence does not contain any termination codons (the termination codon is located at the end of the sequence). The other two reading frames on the same strand are considered "terminating" (TER) reading frames when such termination codons are present within the sequence. In the shifted states, completely different sequences of codons are obtained, and each of these sequences will contain multiple termination codons ("ter"). On the two strands of DNA, theoretically (given that the genetic code is based on triplets of nucleotides), six different sequences can be generated: three on one strand and three on the other. This phenomenon is known as "gene overlap." It was first discovered in viruses. Viral genomes are usually very small; gene overlap was necessary to pack as much information as possible into the relatively short molecules of DNA. The large human genome, in theory, could do without such overlap. However, even in the human genome, although relatively rare, gene overlap does occur [99].

According to the rules of the genetic code, information can be translated from the nucleotide language into the amino acid language. This process can be carried out using computer programs.

As a specific example, let's consider a fragment of the nucleotide sequence from the coding region of the SV40 genome, [137], which is shown in Table 6.8.

Table 6.8

```
CCT CGG CCT CTG AGC TAT TCC AGA AGT AGT GAG GAG GCT TTT TTG GAG
Pro Arg Pro Leu Ser Tyr Ser Arg Ser Ser Glu Glu Ala Phe Leu Glu

GCC TAG GCT TTT GCA AAA AGC TTT GCA AAG ATG GAT AAA GTT TTA AAC
Ala Ter Ala Phe Ala Lys Ser Phe Ala Lys Met Asp Lys Val Leu Asn

AGA GAG GAA TCT TTG CAG CTA ATG GAC CTT CTA GGT CTT GAA AGG AGT
Arg Glu Glu Ser Leu Gln Leu Met Asp Leu Leu Gly Leu Glu Arg Ser

GCC TGG GGG AAT ATT CCT CTG ATG AGA AAG GCA TAT
Ala Trp Gly Asn Ile Pro Leu Met Arg Lys Ala Tyr
```

The entire sequence is read from left to right, and from top to bottom. It contains 180 nucleotides (which corresponds to 60 amino acids). As we have already mentioned, one of the main characteristics of DNA is its nucleotide composition. The mathematical model would be much simpler if the bases in the sequence had uniform frequencies.

The analysis of nucleotide sequences has its particularities. In particular, it should be noted that the frequencies of adjacent nucleotides are not independent of each other. For example, the frequencies of pairs of adjacent bases usually differ from the product of the frequencies of those individual bases. Then, if P_{ij} represents the frequency with which adjacent bases occur in a sequence, P_i represents the frequency of base type i, and P_j represents the frequency of base type j, then: $P_{ij} \neq P_i \times P_j$.

For example, in the given nucleotide frequency sequence, the frequencies of nucleotides A and G are equal to each other, both being 0.278. Therefore, the sequences PA ×PA and PA ×PG both equal 0.077. The frequencies of their pairs in the sequence are: P_{AA} = 0.1 and PAG = 0.11, which corresponds to the statement made earlier.

$P_{ij} \neq P_i \times P_j$, because $P_{AA}(0.1) \neq PA \times P_A(0.077)$, and $PAG(0.11) \neq PA \times PG(0.077)$.

Table 6.9 shows the frequencies of the four bases and 16 dinucleotides. Adjacent bases cannot be independent of each other for biological reasons; the connection between adjacent nucleotides is maintained in a population over very long periods of time.

Table 6.9

Nucleotide pairs	**Distribution of nucleotide pairs**		**Counts of bases**
	Number	**Percentage**	
AA	18	10.1	Adenine 50 (27.8)
AT	10	5.6	Guanine 50 (27.8)
AC	2	1.1	Thymine 50 (27.8)
AG	20	11.2	Cytosine 30 (16.6)
CA	6	3.4	
CC	7	4.0	
CG	1	0.5	

Nucleotide pairs	Distribution of nucleotide pairs		Counts of bases
	Number	Percentage	
CT	16	9.0	
GA	17	9.6	
GC	12	6.7	
GG	16	9.0	
GT	5	2.8	
TA	9	5.1	
TC	8	4.5	
TG	13	7.3	
TT	18	10.1	

Table 6.10 shows the number of different codons that occur in the selected sequence (from Table 6.8), relative to all 64 possible triplets of nucleotides and their corresponding amino acids.

Table 6.10

The number of different codons present in the sequence (Reference: Table 6.8)

AAA Lys 2	GAA Glu 2	TAA Stop 0	TGA Stop 0
AAG Lys 2	GAG Glu 4	TAG Stop 1	TGG Trp 1
AAT Asn 1	GAT Asp 1	TAT Tyr 2	TGT Cys 0
AAC Asn 1	GAC Asp 1	TAC Tyr 0	TGC Cys 0
AGA Arg 3	GGA Gly 0	TTA Leu 1	CAA Gln 0
AGG Arg 1	GGG Gly 1	TTG Leu 2	CAG Gln 1
AGT Ser 3	GGT Gly 1	TTT Phe 3	CAT His 0
AGC Ser 2	GGC Gly 0	TTC Phe 0	CAC His 0
ATA Ile 0	GCA Ala 3	GTA Val 0	CTA Leu 2
ATG Met 3	GCG Ala 0	GTG Val 0	CTG Leu 2
ATT Ile 1	GCT Ala 2	GTT Val 1	CTT Leu 2
ATC Ile 0	GCC Ala 2	GTC Val 0	CTC Leu 0
ACA Thr 0	CGA Arg 0	TCA Ser 0	CCA Pro 0
ACG Thr 0	CGG Arg 1	TCG Ser 0	CCG Pro 0
ACT Thr 0	CGT Arg 0	TCT Ser 1	CCT Pro 3
ACC Thr 0	CGC Arg 0	TCC Ser 1	CCC Pro 0

An efficient algorithm for finding straight repetitions in a sequence of bases was proposed by Samuel Karlin and others (Karlin et al., 1983) [138]. Throughout this process, the entire sequence is being examined, and "words" are generated using specific sets of "letter" bases (i.e., the underlying characters from which the words are composed). The letters of the alphabet are assigned numbers as follows: A = 0, G = 1, T = 2, C = 3, and so on. Each sequence or word from the set of strings consisting of k letters, such as X_1, X_2, …, X_k, is assigned a unique value, which is determined by the following rule:

$1 + \sum_{i=1}^{k} \alpha_i 4^{k-1}$. (6.34)

This value falls within the range of 1 to 4k. A five-letter word such as "GCAAA" has a corresponding value.

$1{+}1 \times 4^4{+}3{\times}\ 4^3{+}0 \times 4^2{+}0 \times 4^1{+}0 \times 4^0 = 1{+}1 \times 256 + 3 \times 64 = 449.$

Consider a sequence that consists, for example, of 20 letters:
GCAAAAAGCTTTGCAAAGAT

The search for direct repetitions of length greater than two bases begins with the examination of two-letter repetitions. The two-letter words, with all their 16 possible combinations, are located in the corresponding positions as shown in Table 6.11.

Table 6.11

Value	Position	Value	Position
1 (AA)	3, 4, 5, 6, 15, 16	9 (TA)	—
2 (AG)	7, 17	10 (TG)	12
3 (AT)	19	11 (TT)	10, 11
4 (AC)	—	12 (TC)	—
5 (GA)	18	13 (CA)	2, 14
6 (GG)	—	14 (CG)	—
7 (GT)	—	15 (CT)	9
8 (GC)	1, 8, 13	16 (CC)	—

The words with the values 1, 2, 8, 11, 13 are repeated here.

The meanings of the three-letter words that start with a repetition of two-letter words, as well as their positions within this sequence of 20 letters, are presented in Table 6.12.

Table 6.12

Value	**Position**
1 (AAA)	3, 4, 5, 15
2 (AAG)	6, 16
5 (AGA)	17
8 (AGC)	7
29 (GCA)	1, 13
31 (GCT)	8
42 (TTG)	11
43 (TTT)	10
49 (CAA)	2, 14

The following text contains repetitions of words with specific values (1, 2, 29, 49). Among the three-letter words, those with the value "1" appear most frequently and are located at positions 3, 4, 5, and 15.

Under the assumption of complete independence among nucleotides (i.e., a Markov chain of order 0), asymptotic expressions for the likelihood of the longest repeated sequence of length n were derived. The expected length of such a sequence is equal to:

$$L = \frac{0.6359 + 2\,ln\,n + ln(1-P)}{ln(1/P)} \tag{6.35}$$

It is proportional to the logarithm of the length of the sequence, whereas the variance does not depend on the length of the sequence.

Here, P represents the sum of the squares of the frequencies of the bases in the sequence, which is equal to:

$$P = \sum_{i=1}^{4} p_i^{\,2} \tag{6.36}$$

The calculations can be simplified if you start implementing the Karlin algorithm with a word of length k whose value is closest to the expected average number of occurrences for the longest possible substring.

During structural analysis, it is possible to identify regions of the chain that are rich in purines, pyrimidines, or combinations thereof. If necessary, one can detect specific characteristics in the sequence being analyzed, including direct and inverted repeats of nucleotides or amino acids. Information about the homology or partial homology of various sequences can also be obtained. Such comparative analysis is widely used in the study of evolution at the molecular level. To compare two sequences with each other, or a single sequence with a group of other sequences, the probabilistic analysis described above, utilizing a Bayesian recognition procedure, can be applied.

Comparing different sequences can be done in various ways. The simplest method is a visual approach using dot diagrams. The sequences to be compared are placed in the cells of a table. A dot is placed at the intersection of rows and columns that contain the same letter. If dots are arranged diagonally, it indicates the presence of identical sub-sequences within the sequences. To reduce the number of dots that may result from random coincidences, a filtering method is used: dots are only assigned when a certain proportion of bases match in the shorter sequences. Areas of high similarity between two sequences in the diagram are represented by diagonal lines with a large number of dots, indicating k matching bases. Even if two sequences are homologous (i.e., related by evolution), their nucleotide composition and length are unlikely to remain unchanged. Therefore, some discrepancies are allowed when detecting local similarities between sequences. When searching for matches, all possible starting positions are considered; if a match is found at least at the first few bases, the corresponding region is expanded. If the bases in the next position of both sequences match, they are added to the expanding region. However, a position with non-matching bases may also be included if a certain number of subsequent matches are satisfied (for example, two out of three bases must match). If this condition is not met, a certain number of bases are removed from one of the amino acid sequences, and the search continues from that point. This removal of one or two bases is only allowed if two or three out of the next three bases match. Finally, sequences with non-matching bases can be added to the expanding region if the similarity between the resulting sequences exceeds a predetermined threshold. Applying this algorithm to the short sequences “CAAAGATTC” and “CAAAAAAGC” yields the following longest matching regions:

CAAAGA
CAAAAA.

In this sequence, groups of four nucleotides (CAAA/CAAA) are followed by non-matching bases (G/A), which are then again replaced by matching bases. The algorithm allows for exactly one mismatch, one insertion, or one deletion; in the subsequent three positions, there must be at least two matching bases.

It can be assumed that the analysis of protein sequences is more informative than the analysis of DNA sequences. In this case, there are no issues related to the degeneracy of the genetic code. Since protein sequences are represented using an alphabet of 20 letters, it is possible for two proteins to have the same amino acid at a given position by chance; however, this probability is lower than the probability of two nucleotide sequences having the same nucleotide at that position. To analyze amino acid sequences, more sophisticated methods are required, which take into account the differences in the degree of similarity between amino acids and the resulting variations in the likelihood of one amino acid evolving into another. For example, it is useful to classify amino acids into the following groups: non-polar (hydrophobic): G, A, V, L, I, F, P, M; polar (hydrophilic): S, T, Y, W, N, C, Q; basic: K, R, H; and acidic: D, E. Other classification methods are also possible.

Contact maps are used to simplify the representation of a protein’s three-dimensional structure. They contain information only about the distances between amino acids in three-dimensional space and are represented as a binary symmetric square matrix M with dimensions $N \times N$. In this matrix, the element $M(i, j)$ equals 1 if the distance between the i-th and j-th amino acid residue in the protein is less than a certain threshold value; otherwise, $M(i, j)$ equals 0. Here, $i = 1,\ldots, N$ and $j = 1,\ldots, N$ represent the ordinal numbers of the amino acids in the protein’s primary structure, and N denotes the total number of amino acids in the protein.

Many different methods have been developed for predicting the contact map of a protein based on its known primary structure. These methods can be divided into two categories: 1) statistical approaches that utilize correlation mutation techniques; and 2) approaches that employ machine learning algorithms. The accuracy of contact map predictions depends on the specific method used, as well as the quality of the multiple sequence alignment and the precision of the secondary structure predictions. Two sequences are considered to be aligned when they are arranged in a way that maximizes their similarity to each other.

The general task of calculating the coordinates of protein atoms in three-dimensional space that are compatible with a given contact map is known as the “existence of a graph with disks of unit radius” problem; this problem is classified as an NP-hard task [10].

6.5. PROTEIN STRUCTURES AND METHODS FOR THEIR ANALYSIS

The central proposition of molecular genetics, which outlines the pathway by which genetic information is expressed within a cell, is: DNA → RNA → Protein.

Each cell consists of approximately 100,000 proteins. Thanks to the flexible bonds (peptide bonds) between amino acids, these proteins can assume numerous different shapes. The specific shape of a protein depends first on its physical structure, which is determined by the sequence of amino acids, and secondly on the interaction between the positive and negative charges at the ends of the connected amino acids. As a result of this interaction, amino acids behave like magnets.

In 1959, the Danish biochemist K. Lindström proposed distinguishing between four levels of protein structural organization:

1. The primary structure of a protein represents the order of amino acids in the polypeptide chain.
2. As a secondary structure, any spatial arrangement of the segments of a polypeptide chain is considered; particular attention is paid to regular spatial structures. The most common types are α-helices and β-sheet folds.
3. Tertiary structure: This refers to the spatial arrangement of all the atoms in the polypeptide chain that makes up a protein molecule. In most cases, the polypeptide chains of proteins fold into compact structures known as globular proteins. Within these globular proteins, a hydrophobic core is formed consisting of amino acid residues such as Ala (Alanine), Val (Valine), Ile (Isoleucine), Leu (Leucine), Met (Methionine), and Phe (Phenylalanine).
4. Quaternary structure: This refers to the way in which subunits (polypeptide chains) are arranged in space, connected to each other by non-covalent interactions to form a protein complex (Figure 6.1) [125].

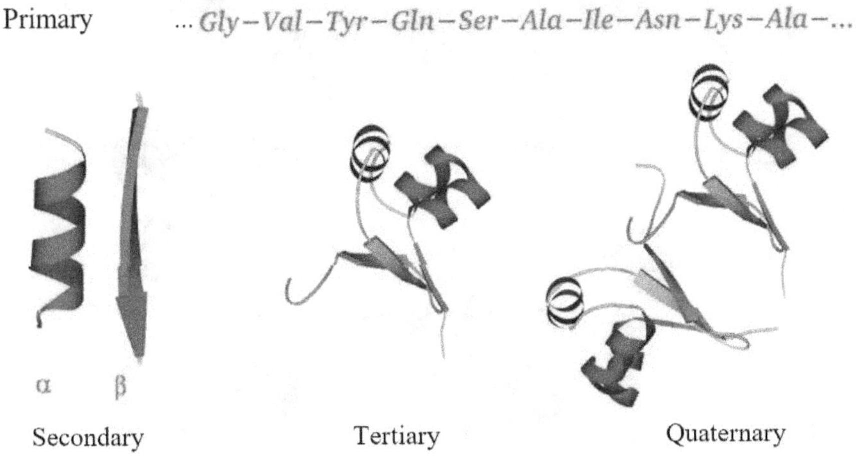

Figure 6.1. Structures of protein organization

It is known that the structure of a protein is determined by the sequence of amino acids. However, no one has yet been able to accurately predict the structure of a protein based solely on its amino acid sequence. The process by which a protein molecule folds, and why it folds in a particular way rather than another, remains incomprehensible to this day. Perhaps precise sciences such as mathematics and bioinformatics can help geneticists solve this problem.

Two types of protein polypeptide chain folding patterns—alpha-helices (helical segments) and beta-structures (sheet-like or layered folds)—were discovered by L. Pauling and his colleagues in 1951.

The alpha helix is the most stable conformation of a protein. One of the factors that stabilize the alpha helix is the formation of a hydrogen bond between two HN–C=O groups, as in [104].

The right-handed α-helix is the most common type of secondary structural element in proteins. Each turn of the helix consists of 3.6 amino acid residues, and the pitch of the helix (i.e., the minimum distance between two adjacent equivalent points) is 0.54 nm. The α-helix is stabilized by nearly linear hydrogen bonds formed between the NH groups of one amino acid residue and the CO group of the fourth amino acid residue in the sequence. As a result, in longer helical regions, each amino acid residue contributes to the formation of two hydrogen bonds. These conformationally rigid segments often consist of 4 or 7 amino acid residues, which corresponds to specific helix pitches [Shataev L.K. et al., 2006]. Figure 6.2 shows an α-helix in a cross-sectional view; the side chains are located on the outside of the helical structure.

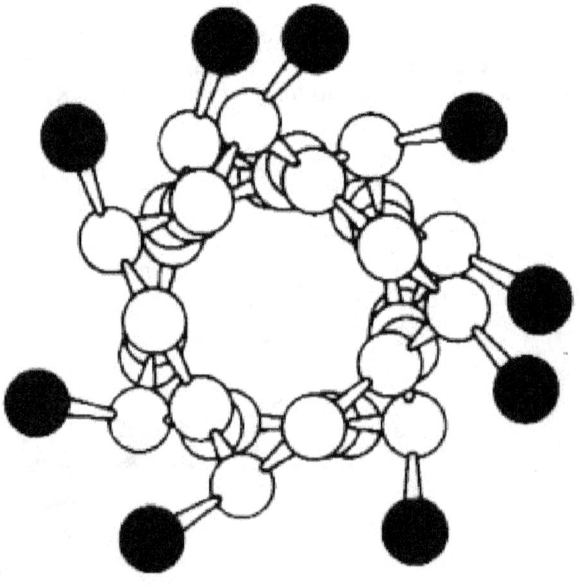

Figure 6.2. Cross-sectional view of a spiral

This type of structure can be disrupted by certain amino acids, especially glycine and proline.

Some polypeptides adopt an alpha-helix structure in solution, while others do not exhibit this ability. Examples of amino acids that form alpha-helices include alanine and leucine; those that do not include serine and threonine. Knowing the structure of the alpha-helix backbone, one can identify the segments of the amino acid sequence that are capable of forming alpha-helices, based on these characteristics (sometimes supplemented by requirements for specific residue sequences). Statistical analyses can also be used to determine the distribution of both alpha- and beta-structures in various combinations of amino acids. For instance, the prevalence of glycine and proline residues in regular sequences (which are essential for the bending of the peptide chain) facilitates the transition of macromolecules into a superhelical conformation, in which three or seven alpha-helix segments twist together to form a structure resembling a three-stranded or seven-stranded cable.

M.V. Volkenstein [121] grouped amino acids with similar structures:

1. Amino acids containing hydrocarbon groups:

Glycine (Gly); Alanine (Ala); Valine (Val); Leucine (Leu); Isoleucine (Ile); Phenylalanine (Phe); Proline (Pro); Tryptophan (Trp).

2. Amino acids containing sulfur: methionine (Met); cystine (Cys).
3. Amino acids containing a hydroxyl group are: Threonine (Thr); Serine (Ser); Tyrosine (Tyr).
4. Amino acids containing an additional acidic group:

Asparaginic acid (Asp); Glutamic acid (Glu).

5. Amino acids containing an amide group:

Asparagine (Asn); Glutamine (Gln).

6. Amino acids containing a basic group:

Histidine (His); Lysine (Lys); Arginine (Arg).

Based on the ability of amino acids to form alpha-helices in proteins, they are classified into those with high, medium, or low capacity for doing so. Amino acids with high capacity include: leucine, methionine, histidine, glutamine, lysine, glutamic acid, cystine, and alanine. Amino acids with medium capacity include: phenylalanine, isoleucine, valine, asparagine, tryptophan, arginine, and aspartic acid. Amino acids with low capacity include: serine, proline, threonine, glycine, and tyrosine.

The grouping of amino acids according to their preference for forming certain structures is shown in Table 6.13:

A measure of these most important amino acid properties, which contribute to the structural diversity of proteins, can be represented by the relative frequency of these amino acids in various elements of the protein's secondary structure. This information is provided in [Berg et al., Biochemistry, 2002, pp. 123–124].

Arginine does not show a clear preference for any particular structure.

After the translation of the amino acid sequence from an RNA molecule is completed, these amino acids begin to fold into a three-dimensional structure. This process results in the formation of a new chemical compound: a protein. For a protein to function properly, it requires DNA of the correct configuration. For example, in the case of Alzheimer's disease, the DNA code is damaged, and the protein is believed to transform into a parasitic-like structure. Many factors contribute to the development of Alzheimer's disease; however, there are also genetic defects that affect the normal functioning of cells. This disease is currently incurable because genetic disorders are difficult to treat. Nevertheless, this is merely a matter of the insufficient advancement of bioengineering technology at present. Scientists believe that over time, experts will learn how to replace malfunctioning genes with the correct ones (e.g., [139]).

Table 6.13

№	Amino Acid	α-helix	β-sheet	β-turn
1	Ala	0.434	0.303	0.263
2	Cys	0.42	0.28	0.30
3	Leu	0.45	0.35	0.2
4	Met	0.52	0.34	0.14
5	Glu	0.45	0.24	0.31
6	Gln	0.42	0.26	0.32
7	His	0.41	0.36	0.23
8	Lys	0.42	0.26	0.32
9	Val	0.32	0.52	0.16
10	Ile	0.33	0.5	0.17
11	Phe	0.36	0.44	0.2
12	Tyr	0.24	0.41	0.35
13	Trp	0.34	0.4	0.26
14	Thr	0.27	0.39	0.34
15	Gly	0.18	0.29	0.53
16	Ser	0.26	0.31	0.43
17	Asp	0.33	0.23	0.44
18	Asn	0.31	0.26	0.43
19	Pro	0.17	0.21	0.62
20	Arg	0.32	0.34	0.34

In the early 1950s, F. Sanger showed that the protein chain has a unique sequence of units — amino acid residues (an "amino acid residue" refers to what remains from a free amino acid after it is incorporated into the protein chain). The protein chain consists of a chemically regular backbone, from which various side chains of amino acids extend, for example: R_1, R_2, …, R_m.

```
–NH – CH – CO – NH – CH – CO – ... – NH – CH – CO –
      |               |                    |
      R1              R2                   Rm
```

The positions of all 20 standard (proteinogenic) amino acid residues in a protein chain are encoded by the gene.

Based on their general structural type, proteins can be divided into three classes: fibrous proteins; membrane proteins; and water-soluble, globular proteins that reside in water.

The vast majority of three-dimensional protein structures belong to water-soluble globular proteins. For membrane-bound and fibrous proteins, only a few spatial structures or individual fragments have been elucidated to date. The secondary structure of proteins is characterized by a regular, periodic arrangement (the "conformation") of the main polypeptide chain, although the conformations of their side chains vary greatly. The structure of a protein is determined by its amino acid sequence. In other words, the amino acid sequence determines the protein's spatial structure, and the protein's structure in turn determines its function—i.e., with which other molecules it interacts and what processes occur as a result of those interactions.

Amino acids, as fragments of the polypeptide chain, are presented in Appendix 2.

The protein chain, consisting of twenty standard amino acid residues, originating from the Cα atoms of the main chain, is shown in Figure 6.3. White circles represent the Cα atoms of the main chain; gray circles represent the carbon (C) atoms of the side chains; light-gray circles represent the hydrogen (H) atoms of the side chains; blue circles represent the nitrogen (N) atoms of the side chains; red circles represent the oxygen (O) atoms of the side chains; yellow circles represent the sulfur (S) atom of the side chain [125].

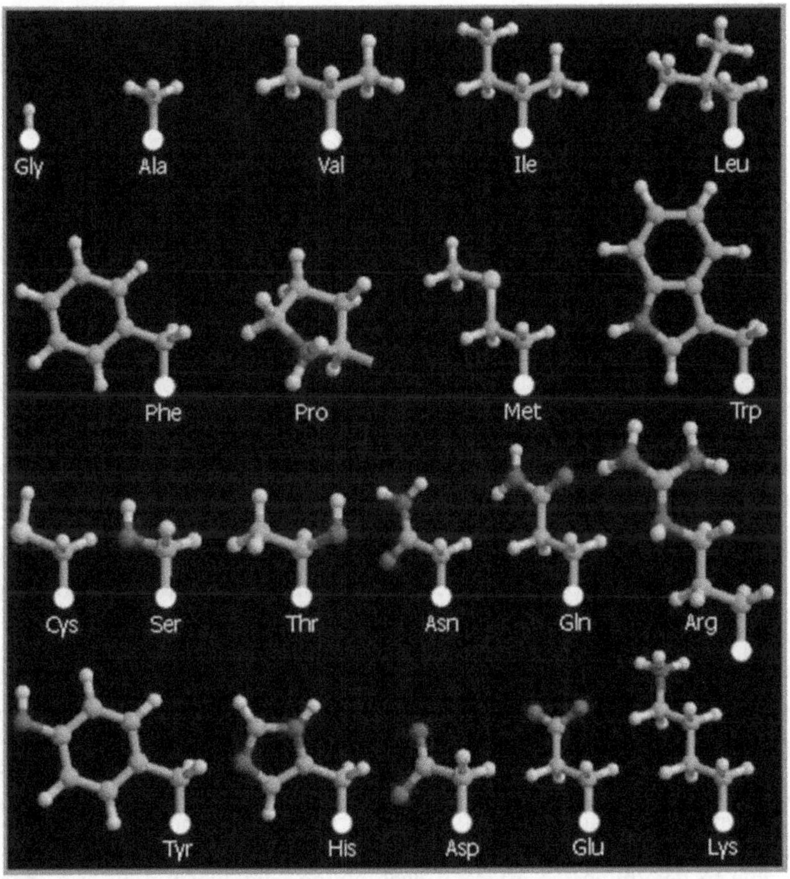

Figure 6.3. Side chains of the twenty standard amino acid residues originating from the Cα atoms of the main chain

The side chains can be divided into the following groups:

1. Polar side chains are formed by thc following elements: Asn, Gln, Asp, Glu, Arg, His, Lys, Ser, Thr.
2. The side chains have weakly defined polarities. They are formed by: Tyr, Cys, Trp.
3. The neutral side chains are formed by Gly.
4. Non-polar side chains are formed by the following elements: Ala, Ile, Leu, Pro, Val, Phe, Met.

These groups can be formed based on the first two nucleotides of an amino acid.

They can then be presented in the form of Table 6.14 as follows:

Table 6.14

First\Second	Second Nucleotide			
	A	G	T	C
A	1	1	4	1
G	1	3	4	4
T	2	2	4	1
C	1	1	4	4

In the formation of the stable three-dimensional structure of a protein, which is called a globule, the internal interactions between the amino acids within the protein chain as well as the interactions with molecules in the surrounding environment play a crucial role. Proteins generally perform their functions while being in water. Water molecules are closely bound to each other; this type of bond is known as a hydrogen bond (the bond between an oxygen (O) atom and a hydrogen (H) atom in a molecule, for example, H_2O). The concept of hydrogen bonds was discussed in Section 6.1. Hydrogen bonds are not only found in water; they occur whenever a hydrogen atom is chemically bonded to an electronegative atom (i.e., an atom that attracts electrons) and simultaneously comes into close proximity to another electronegative atom. A water molecule is polar, which means that its atoms have small, partial electric charges: the oxygen (O) atom carries a negative charge, while the hydrogen (H) atom carries a positive charge. Overall, the water molecule has a net charge of zero. The energy associated with a hydrogen bond is approximately 5 kcal/mol. Similar to water, the backbone of a protein chain is also polar; more precisely, the peptide groups that make up the protein chain are polar in nature.

In an alpha-helix, each turn consists of approximately four amino acid residues. Therefore, the residues with numbers 1 and 4 are located next to each other and are connected by a hydrogen bond. The neighboring residues relative to residue 4 (residues 3 and 5) are also very close to residue 1 (at a distance of 2Å).

If there are n residues in a helix, then they are held together by $n-2$ hydrogen bonds. By using short peptides with a predetermined primary structure, it is possible to determine the effect of each amino acid substitution at a specific position on the helical nature of the peptide; in other words, this effect actually depends on the position of the amino acid residue relative to the N- and C-termini of the helix. Similar approaches are also used to measure the stability of the β-sheet structure in polypeptides. In the case of a β-sheet structure (also known as a folded sheet structure), the polypeptide chains are stretched and arranged parallel to each other, and they are connected by hydrogen bonds. Proteins contain two types of β-structures: those with chains oriented in the same direction (parallel beta-structures) and those with chains oriented in opposite directions (antiparallel beta-structures). Beta-structures can be further classified into beta-sheets and beta-turns. A beta-turn causes the polypeptide chain to bend by approximately 180 degrees over a segment containing four amino acid residues; the first and fourth residues in this segment are connected by a hydrogen bond.

The portion of the chain that lacks a defined structure represents an "uncertain" region. It is suggested to model proteins using locally interacting particles, with the interactions themselves being determined algorithmically. This approach allows for the study of cellular processes without relying on the specific nature of protein-protein interactions. Additional possibilities for modeling cellular processes are provided by utilizing potentials that can be used to specify the probabilities of various configurations of the [140] particles.

A protein molecule can assume a very large number of different spatial configurations. The probability that the molecule remains in a particular configuration is higher the lower the energy associated with that configuration. The mechanism by which proteins are folded is not yet fully understood. This problem is known as the "Levinthal Paradox": "The time it takes for a polypeptide to reach its folded state is much shorter than if the polypeptide were to simply explore all possible configurations." How does a protein select its native structure from an infinite number of possibilities? For a chain of 100 amino acids, the number of possible conformations is approximately 10^{100}, and exhaustively exploring them would take approximately 10^{80} years, assuming each transition between configurations takes 10^{-13} seconds. This problem cannot be solved experimentally, as we would have to wait for 10^{80} years. Solution to the paradox: A protein can fold not all at once, but rather by gradually growing a more compact globule through the successive addition of new segments from the protein chain [141].

Amino acids are the building blocks for protein biosynthesis, and in order to use these amino acids to form a protein structure, they must be supplied with the necessary energy. The supply of energy to amino acids is carried out by a substance called adenosine triphosphate (ATP). In all living organisms, ATP serves as a universal energy storage medium. [46] The amino acid becomes activated and interacts with tRNA (transfer RNA), which then transports the amino acid to the sites where the protein molecule is being synthesized. The rest of the text describes the intermolecular forces that are at play within the protein. These forces, which either bend or repel different parts of the protein molecule's structure, can be categorized as physical interactions (or phenomena). The flexibility of the protein molecule allows its various parts to rotate around certain bonds that contribute to the formation of its specific structure. In various structures, the internal forces at play within the molecule balance each other out, resulting in a state that corresponds to the minimum amount of energy. The primary structure of a protein, namely its amino acid sequence, is determined experimentally in a relatively straightforward manner. The DNA sequences of many genes with similar functions (homologous genes), as well as the amino acid sequences of the protein products produced by these genes, are similar across different organisms—ranging from bacteria to higher plants and animals. For example, the cytochrome C protein is well-known for its important role in the production of cellular energy (as mentioned earlier, the energy-carrying molecule is called ATP) through the "burning" of food molecules absorbed by the living cell. Comparing the nucleotide sequences in DNA (or the amino acid sequences of the cytochrome C gene) from different species has shown that this gene is virtually identical across all species that use ATP [27] (with the exception of a few rare mutations). The problem of determining the secondary structure of a protein from the perspective of classical statistical physics boils down to the need to find the global minimum of the free energy for a very complex system – that is, for the protein with a given primary structure in a specific environment. However, applying this approach to determine the structure of a protein molecule is quite complicated. The essence of the problem is as follows: The basis for the overwhelming majority of this type of prediction lies in models that consider the fixed charges of atoms. However, when the molecules interact with a solvent, dipoles are formed on the atoms due to quantum effects, which ultimately affect the accuracy of the protein structure prediction. Furthermore, the principal difficulty here lies in the astronomically large number of possible protein conformations, which makes it not only practically impossible to find the global minimum of free energy, but also fundamentally impossible to accomplish the task [142].

Predicting the secondary structure of a protein based on its primary structure is one of the necessary steps in predicting the tertiary structure and functional properties of new proteins.

The observation that each protein spontaneously folds into a unique three-dimensional native conformation leads to the idea that nature must possess some kind of algorithm for predicting the protein's spatial structure based on its amino acid sequence. Attempts have been made to understand this algorithm in order to create an efficient computer program that would address this highly important yet simultaneously complex problem.

There are a great many methods for determining the structure and composition of molecules. The primary structure of a protein, namely its amino acid sequence, can be determined relatively easily through experimental means. However, determining the secondary structure of a protein involves significant complexities. X-ray crystallography is one of the commonly used techniques for this purpose. There is also an experimental method based on nuclear magnetic resonance (NMR), which can detect alpha-helical or beta-sheet structural elements in proteins by measuring the chemical shifts of atomic nuclei within the molecule; these shifts provide information about the presence and location of such structural regions. The drawbacks of these experimental methods include low accuracy, as well as the complexity and high cost of the technical procedures required. Additionally, there are other methods for predicting the secondary structure of proteins based on their primary structure. Nevertheless, these methods have limitations such as the potential to predict incorrect secondary structural elements or the failure to identify all relevant regions. Moreover, they can only distinguish between helical regions and other types of protein structures; this is a rather crude prediction, given that there are various types of protein structures in existence, including alpha-helices, beta-helices, and numerous other specific structural motifs [117].

The document [143] provides an overview of current approaches to predicting the spatial structure of proteins. According to the authors, the most promising methods are those based on the use of probabilistic models (such as Markov chain models, conditional random fields, etc.).

One such method is the use of Markov chains. The application of mathematical techniques enables the exclusion of numerous secondary factors that can distort the true picture, even in cases where experiments are very precise and quantitative. The method under study belongs to the field of machine learning and operates on training datasets. Open databases containing experimentally determined secondary protein structures can be used as these training datasets. Observations have shown that homologies between two natural proteins with more than 30–35% sequence similarity can be reliably identified (with the caveat that this is only true for sequences longer than 50 residues, preferably 100 residues). However, it should be noted that a 30–35% similarity score, while generally indicating a genetic relationship between proteins, only allows for an accurate overlap of approximately 70–80% of their spatial structures. As a result, the similarity of the remaining 20–30% of the protein structures is inaccurately represented. To accurately determine the structure of 95% of the "new" protein's primary chain, the homology score with a protein of known structure must reach 40–50%.

All the patterns observed in protein structures are of a probabilistic, frequency-based nature; therefore, a Bayesian procedure is used to predict the three-dimensional structure of proteins. The vast majority of experimental and statistical findings can be easily understood in terms of the physics and stereochemistry of amino acids. For instance, Proline (Pro) tends not to be part of an α-helix (except for the N-terminus region of the helix) nor of a β-sheet structure, as it lacks an NH group and is unable to form the hydrogen bonds characteristic of these structures. Moreover, at the N-terminus of the helix the NH group is not required to form such bonds, which is why Proline is frequently found in this position. Alanine (Ala), on the other hand, stabilizes α-helices, while Glycine (Gly) destabilizes both α-helices and β-sheets. Branching amino acids like Valine (Val), Isoleucine (Ile), and Threonine (Thr) further stabilize β-sheets. Hydrophobic amino groups generally have a tendency to be incorporated into α- and β-sheets more frequently. By considering which amino acid residues stabilize different parts of the helix (the N-terminus, the middle, or the C-terminus), we can derive a "template" for helical structures. The template for an α-helix can be described as follows: If the beginning of a protein chain segment is rich in negatively charged groups and contains Proline, if the middle of the segment is enriched with Alanine, Leucine (Leu), and Methionine (Met) residues but lacks Proline and Glycine, and if the C-terminus contains positively charged side chains, then this segment likely represents an α-helix. If an amino acid sequence conforms to this template, it is more likely that a helix is present in that location of the chain [125].

Algorithms for predicting the structure of proteins

As a reminder, an algorithm is an intuitively understandable concept, denoted by [29]. It represents a program written in any universal programming language. An algorithm does not require that the calculations performed by it be absolutely correct; it is possible to write a program that contains potential errors. Therefore, much of the outcome depends on how the program is written and on the skills of the programmer itself.

The Bayesian approach, as mentioned earlier, is based on the use of formula (6.30).

The Bayesian recognition procedure belongs to the field of machine learning methods and operates on training datasets. The training datasets used consist of open databases containing experimentally determined secondary protein structures, which are identified based on protein homology. In other words, the structures of homologous proteins serve as the training data for the Bayesian recognition procedure.

Model building based on homology is a method for predicting the structure of a protein given its sequence, assuming that the protein under study is related to at least one other protein for which both the sequence and the structure are known. The approach relies on the premise that sequences with high similarity are likely to have similar structures. Therefore, if we find good structural candidates among one or several homologous proteins, we have a higher likelihood of identifying the appropriate structure for our target sequence. The quality of the resulting model depends on the degree of sequence similarity between the homologous proteins.

The first step in homology modeling is to align the amino acid sequences of the protein of interest with those of a protein (or proteins) that have a known structure. In most cases, insertions and deletions will occur in the regions of the sequence that comprise the loops between alpha helices and beta strands.

More detailed subsequent steps for modeling based on homology are described in [117].

Next, using a Bayesian recognition procedure on the training datasets, we construct a table that represents the probabilistic preferences for various amino acid pair combinations (see Table 6.15). In this case, Table 6.15 only illustrates the general principle behind the construction of such a table; the complete table is not provided, as the probability values for specific protein structures can vary for different proteins.

The table is generated for all possible pairs of amino acids that are present in the amino acid sequence being studied. The maximum number of possible pairs is $20 \times 20 = 400$.

The sequence in question is then broken down into segments. Each segment typically consists of five amino acids; however, depending on the alignment of the sequences, in some cases the segments may contain between three and ten amino acids.

The choice of the five-amino-acid segment is based on the fact that a fragment of the protein's primary structure consisting of these five amino acids allows for the identification of the starting point of the protein helix. Once this starting point is known, by analyzing the combinations of the individual amino acids in that region, the possible states of the protein's structure can be determined.

The cutting of primary protein structures into fragments of five amino acids corresponds to the maximum number of hydrogen bonds in the core of the helix.

Table 6.15

Probable preference structures for amino acid pairs

	Asn	**Arg**	**...**	**...**	**Ser**
Lys	Rα =0.365; Rβc = 0.26; Rβi =0.375	Rα =0.37; Rβc=0.3; Rβi =0.33	...	...	Rα = 0.34; Rβc=0.285; Rβi=0.375
....	....	....	...	...	....
Pro	Rα =0.24; Rβc =0.235; Rβi =0.525	Rα =0.245; Rβc=0.275; Rβi=0.48	...	...	Rα = 0.215; Rβc=0.26; Rβi =0.525

If a selected fragment consisting of five amino acids contains four consecutive amino acids that have a tendency to form helices, then the entire fragment can be considered helical as well. The helical formation continues towards the C-terminus until it is inhibited by two adjacent residues that have a tendency to cause the protein to fold in a disordered manner. Similar arguments apply to the structures β_s and β_i.

After training on sequences of homologous proteins with known structures, the Bayesian recognition system is ready to predict the structure of the protein under study. A fragment consisting of five amino acids is selected from the sequence of the protein under study.

.... a_1 a_2 a_3 a_4 a_5....

	a_1	
First step: .... a_3 a_4	a_2	a_5......
Second step: .... a_1 a_4 a_5	a_3	....
Third step: .... a_1 a_2 a_5	a_4	....
Fourth step: .... a_1 a_2 a_3	a_5	......

At each step of the process, a pair of amino acids is fed into the Bayesian formula-based algorithm. As a result of the algorithm's processing to identify the secondary structure according to formula 6.38, probability estimates are determined for the current amino acid a_2 from the pairs $a1a_2$ and a_2a_3; for the current amino acid a_3 from the pairs a_2a_3 and a_3a_4; and for the current amino acid a_4 from the pairs a_3a_4 and a_4a_5.

The status of the structure for each current amino acid is determined by the following formula:

$$a_i = \operatorname{argmax}\{P(\alpha), P(\beta_c), P(\beta_i)\}.$$

In most cases, the algorithms for modeling the spatial structure of a protein are derived from its amino acid sequence.

Recently, attempts have been made to determine the genetic code for the spatial structure of proteins by analogy with the genetic code for their amino acid sequences, based on the analysis of experimental data and physical modeling results. However, these efforts once again focused on characterizing the amino acid residues in terms of the first two nucleotides of a codon, with the second nucleotide of the triplet playing a decisive role (see section 6.1). V.V. Sokolik [147] argues that the third nucleotide of a codon does not determine the amino acid residue in the polypeptide chain; rather, it codes for the configuration of the peptide bond in the combination with adjacent amino acid residues. This fact highlights the importance of the third nucleotide: it primarily interacts with the first nucleotide of the tRNA anticodon to form a Crick-like pair, and ultimately determines the overall structure of the protein's spatial arrangement.

In [146], a matrix representing the topological genetic code is presented. In this code, the second nucleotide plays a key role in determining whether an amino acid will exhibit a bonding or anti-bonding behavior: pyrimidines (C and U) are responsible for anti-bonding operations, while purines (A and G) are responsible for bonding operations. Under the term "topological code," V.A. Karasev refers to the topological structure of the genetic code, which results from the objective hierarchy of the L-structure of the six-dimensional Boolean hypercube of codons.

The approach described in the work [147], as noted by V.V. Sokolik, should serve as the foundation for modeling proteins based on their determining nucleotide sequences by constructing a personalized structural template. A well-developed modeling methodology that relies on the amino acid sequences of proteins should then act as an overlay on this foundation, enabling the optimization of the spatial structure of the protein template taking into account the interactions between the side chains of the amino acid residues. Since all the information regarding the conformation of a protein is contained within its gene, it is logical to model the spatial structure of proteins based on their determining nucleotide sequences.

6.6. EXAMPLE OF A BAYESIAN PROCEDURE FOR RECOGNIZING THE SECONDARY STRUCTURE OF PROTEINS (MAKING PREDICTIONS ABOUT THE PROTEIN'S SECONDARY STRUCTURE)

There are two theories regarding the prediction of protein structures. According to the first theory, the protein structure is determined as the result of a kinetic folding process. According to the second theory, it is determined as the structure that has the lowest possible free energy for the given protein chain. Both approaches can lead to the correct outcome, because the most stable structure of a protein chain is formed fairly quickly, despite Levinthal's concerns. Therefore, when predicting protein structures, one can consider the resulting structure to be the stable form of the protein.

In reference [125], the most significant findings obtained using these methods are highlighted. It is noted that α-helices stabilize the amino acids Ala, Leu, and Met; β-strucures stabilize Ile, Val, Thr, Leu, Met, Cys, Phe, Tyr, and Trp; whereas β-folds and loops stabilize Gly and Pro.

It is also important to take into account the evidence that supports the existence of proteins with different functions but very similar structures. This information remains a mystery for experts [104].

Several studies have noted that there currently does not exist a completely reliable and accurate method for predicting the structures of proteins. There are two main reasons for this: 1) the limited accuracy of the energy calculations on which the theoretical prediction of protein structures is based; and 2) the relatively small difference between the structures of proteins that are "correctly" and "incorrectly" folded.

Given the primary structure of a protein, which is the linear sequence of amino acids, it is necessary to assign each amino acid to one of the possible types of secondary structural motifs: alpha-helix, beta-sheet, or beta-turn. As training datasets, open databases containing experimentally determined secondary protein structures are used. We will apply Bayes' formula for the recognition of protein secondary structures in a manner similar to that described in [9] and [10]. To represent the different types of secondary structures, we will use the following notations: "α" for alpha-helices; "β-sheet" and "β-turn" for beta-sheet and beta-turn motifs, respectively. These notations correspond to the values A_1, A_2, A_3 respectively. The recognition process will be based on six diagnostic features: x_1, x_2, x_3, x_4, x_5, x_6. All six features correspond to individual amino acids; however, since each amino acid consists of a sequence of three nucleotides, each nucleotide will be associated with a pair of values representing one of these features. Previously, the following mapping was used to assign nucleotides to binary numbers: A = 00; G = 01; T = 10; C = 11. In this case, the desired outcome is that the measurements of an amino acid match the set of values (features) x_1, x_2, x_3, x_4, x_5, x_6. According to formula (6.28), the probability distribution will satisfy the relevant conditions.

$$P(As \mid a1, a2, \dots, an) = \frac{P(As) \cdot P(a1|As) \cdot P(a2|As) \cdot \dots \cdot P(an|As)}{P(a1,\dots,aj,\dots,an)} \quad (6.37)$$

where a_1, a_2, …, an are sequences of amino acids;

As represents the state of an amino acid.

$P(a_k|A_s)$—these represent conditional probabilities.

$P(A_s)$ – this represents the prior probability of the state of the amino acid.

The number of classes in As is 60; this corresponds to the 20 amino acids and the 3 types of secondary structures. All six features relate to one amino acid, but since each amino acid consists of a sequence of three nucleotides, each nucleotide will correspond to a pair of values representing one feature. For example, an amino acid sequence like "AAA" would be represented as 000000, while "AGC" would be represented as 000111. In other words, the input to the system is an amino acid, and the output is the corresponding secondary structure of that protein. Thus, the primary structure of the protein, along with its respective secondary structure, can be represented in this format.

V	Y	G	S	A	I	N	K	I	H	A	G	R	K	I	F
β_i	β_i	α	α	α	α	α	α	α	α	α	β_i	β_i	β_c	β_c	β_c

where α denotes an alpha-helix;

β_c – This represents the beta layer.

β_i – This represents a beta version of the "beta bend" feature.

The a priori probabilities should be determined based on previous experiments. Suppose they are distributed as follows: $P(A_1) = 0.4$; $P(A_2) = 0.15$; $P(A_3) = 0.45$.

In the training mode, conditional probabilities are determined.

The original data for calculating the conditional probabilities $P(x_j|As)$, which are necessary for determining the corresponding protein structures, are provided in tables 6.16–6.18.

Using the column x_1 from Table 6.16 as an example, we will demonstrate how to calculate conditional probabilities. In the column x_1, out of the five values, two are equal to 1. Therefore, the probability that the feature is present in the element $P(x_1|A_1)$ is 0.4. Consequently, the probability that the feature is absent from the element is…

$$P(\bar{x}1|A_1) = 0.6.$$

Table 6.16

The original data used for calculating conditional probabilities…

The structure "α" = A_1

Amino acid number	Measurements					
	x_1	x_2	x_3	x_4	x_5	x_6
1	0	0	0	1	0	1
2	0	0	0	1	1	1
3	1	0	0	0	1	1
4	1	1	0	0	1	1
5	0	1	1	0	1	1

Table 6.17

The original data used for calculating conditional probabilities…

The structure $\beta_c = A_2$.

The number of the amino acid:	Measurements					
	x_1	x_2	x_3	x_4	x_5	x_6
6	0	0	1	1	1	0
7	0	1	1	1	0	0
8	1	0	1	1	0	1
9	1	1	1	0	0	0
10	0	1	0	1	0	0

Table 6.18

Structure $\beta_l = A_3$

The number of the amino acid:	Measurements					
	x_1	x_2	x_3	x_4	x_5	x_6
11	0	0	0	0	1	0
12	0	0	1	1	1	0
13	1	0	0	1	1	0
14	1	1	1	1	1	0
15	0	1	1	1	0	1

The conditional probabilities for the cases presented in tables 6.16–6.18 are distributed as follows:

$P(x_1|A_1) = 0.4$ $P(\bar{x}1|A_1) = 0.6;$

$P(x_1|A_2) = 0.4$ $P(\bar{x}1|A_2) = 0.6;$

$P(x_2|A_1) = 0.4$ $P(\bar{x}2|A_1) = 0.6;$

$P(x_2|A_2) = 0.6$ $P(\bar{x}2|A_2) = 0.4;$

$P(x_3|A_1) = 0.2$ $P(\bar{x}3|A_1) = 0.8;$

$P(x_3|A_2) = 0.8$ $P(\bar{x}3|A_2) = 0.2;$

$P(x_4|A_1) = 0.4$ $P(\bar{x}4|A_1) = 0.6;$

$P(x_4|A_2) = 0.8$ $P(\bar{x}4|A_2) = 0.2;$

$P(x_5|A_1) = 0.8$ $P(\bar{x}5|A_1) = 0.2;$

$P(x_5|A_2) = 0.2$ $P(\bar{x}5|A_2) = 0.8;$

$P(x_6|A_1) = 1$ $P(\bar{x}6|A_1) = 0;$

$P(x_6|A_2) = 0.2$ $P(\bar{x}6|A_2) = 0.8;$

$P(x_1|A_3) = 0.4$ $P(\bar{x}1|A_3) = 0.6;$

$P(x_2|A_3) = 0.4$ $P(\bar{x}2|A_3) = 0.6;$

$P(x_3|A_3) = 0.6$ $P(\bar{x}3|A_3) = 0.4;$

$P(x_4|A_3) = 0.8$ $P(\bar{x}4|A_3) = 0.2;$

$P(x_5|A_3) = 0.8$ $P(\bar{x}5|A_3) = 0.2;$

$P(x_6|A_3) = 0.2$ $P(\bar{x}6|A_3) = 0.8$

The schematic diagram of the recognition system, which operates using Bayes' formula, is shown in Figure A.3.1 (Appendix 3).

The teacher already knows in advance that the amino acids with numbers 1–5 (Table 6.16) belong to the structure "A_1", and the amino acids with numbers

6–10 (Table 6.17) as well as 11–15 (Table 6.18) belong to the structures "A_2" and "A_3" respectively. This information represents the teacher's "experience", which is stored in their memory. Each of these groups of amino acids (five amino acids in each group) is referred to as a sample. The teacher feeds these samples into the recognition system, and the system calculates all the relevant probabilities and stores them in its memory. In addition, the system's memory also contains the values of the prior probabilities.

The choice of the prior distribution in applied problems of Bayesian estimation is one of the most important tasks. The determination of the prior distribution and all subsequent calculations are carried out within the framework of this procedure (the Bayesian approach).

At this point, it can be said that there are three main approaches for recommending the choice of an appropriate prior distribution. These approaches are based on: 1) the principle of conjugacy; 2) the absence of prior information; 3) the informational criterion [51].

After the system has been trained, it is ready to recognize (predict the secondary structure).

As a predictable secondary structure, any amino acid of the type for which the system has been trained can be chosen.

For example, suppose you need to determine the secondary structure of an amino acid based on a set of parameters: $x_1 = 1$, $x_2 = 1$, $x_3 = 0$; $x_4 = 1$, $x_5 = 0$, $x_6 = 1$. The system performs calculations according to formula (6.37) using the conditional probabilities for these values (x_j) that are stored in memory. The probability value that appears in the denominator of formula (6.37) is also derived from this process.

$$P(a1, \dots, aj, \dots, an) = \sum_{i=1}^{r} \prod_{j=1}^{6} P(As) \cdot P(xj|As)$$

This is called the "full probability." To determine the full probability, we find the numerator (6.37) by matching each amino acid with all of its six corresponding features (where each amino acid corresponds to the set hi, which consists of six features in total).

$P(A_1) \times P(x_1 \mid A_1) \times P(x_2 \mid A_1) \times P(\overline{x}_3 \mid A_1) \times P(x_4 \mid A_1) \times P(\overline{x}_5 \mid A_1) \times P(x_6 \mid A_1) = 0.4 \times 0.4 \times 0.4 \times 0.8 \times 0.4 \times 0.2 \times 1 = 0.0041$;

$P(A_2) \times P(x_1 \mid A_2) \times P(x_2 \mid A_2) \times P(\overline{x}_3 \mid A_2) \times P(x_4 \mid A_2) \times P(\overline{x}_5 \mid A_2) \times P(x_6 \mid A_2) = 0.15 \times 0.4 \times 0.6 \times 0.2 \times 0.8 \times 0.8 \times 0.2 = 0.0009$;

$P(A_3) \times P(x_1 \mid A_3) \times P(x_2 \mid A_3) \times P(\overline{x}_3 \mid A_3) \times P(x_4 \mid A_3) \times P(\overline{x}_5 \mid A_3) \times P(x_6 \mid A_3) = 0.45 \times 0.4 \times 0.4 \times 0.4 \times 0.8 \times 0.2 \times 0.2 = 0.0009$.

Furthermore, we have: $P(a1, \dots, aj, \dots, an) = \sum_{i=1}^{r} \prod_{j=1}^{6} P(xj|As) \cdot P(As)$

$= 0.0041 + 0.0009 + 0.0009 = 0.0059$.

The final result is: $P(A_1 \mid x_1, x_2, \overline{x}3, x_4, \overline{x}5, x_6) = 0.0041 : 0.0059 = 0.7$;

$P(A_2 \mid x_1, x_2, \overline{x}_3, x_4, \overline{x}_5, x_6) = 0.0009 : 0.0059 = 0.15$;

$P(A_3 \mid x_1, x_2, \overline{x}_3, x_4, \overline{x}_5, x_6) = 0.0009 : 0.0059 = 0.15$.

where i (1 – r) represents the number of classes; j (1 – 6) represents the number of features (or characteristics).

By calculating the probabilities for the predicted structure in this manner, the system compares them and indicates the type of secondary structure for which the probability is highest. In our example, the comparator points to the structure A_1.

The calculation of probabilities in the prediction mode is shown in detail for the structure A_1. The probabilities for the structures A_2 and A_3 are calculated in a similar manner.

The in-depth analysis of genetic information is fully justified; indeed, for some proteins, it is possible to establish a nearly deterministic relationship between the protein's secondary structure and its nucleotide sequence.

At the "Nanomir" laboratory, part of the scientific and production association "Picotekhnologiya" affiliated with Moscow State Technical University named after N.E. Bauman, a strong correlation between the spatial structure of a protein and its nucleotide sequence was theoretically predicted using physical modeling methods, experimentally verified, and statistically confirmed. It was found that during the biosynthesis process, the third nucleotide in a codon determines the orientation of the amino acid, thereby forming the specific spatial isomer — that is, the conformation of the protein molecule — and cutting off the competing pathways for the formation of secondary and tertiary structures. [144]

The "Pikotechnology" program takes into account 4 main composition options. Option 1 corresponds to the insertion of the residue into the 4/10 portion of the alpha-helix; Option 2 corresponds to the insertion of the residue into the classical beta-layer; Option 3 corresponds to the insertion of the residue into a beta-turn; and Option 4 corresponds to the insertion of the residue into the 3/10 portion of the alpha-helix. It turned out that the third nucleotide in the codon codes for the configuration of the peptide bond in the compound formed by the adjacent amino acid residues. It interacts with the first nucleotide of the tRNA anticodon, forming a Crick pair, and ultimately determines the overall framework of the protein's three-dimensional structure. Under the compositional code A.YU. Kuleshov understands the correspondence between the third nucleotide of any codon and the meaning of the corresponding genetic code (ranging from 1 to 4, depending on the number of possible nucleotide combinations). These combinations determine the orientation of the amino acid residue that will be attached to the growing polypeptide chain. Experiments have shown that there are more varieties of right-handed and left-handed helices (2/7, 3/10, α, π, etc.) than is dictated by the structural pattern of peptide bonds. [145]. Therefore, the sequence of peptide bonds in a helical structure only determines the direction of the helix (right-handed or left-handed), while the variety of helical structures is determined by the properties of the amino acid residues that make up the protein chain and form the helical coil. It follows therefore that, based on the structural template of a right-handed helix, through the interaction of the side chains of amino acid residues, a right-handed 3/10 helix, a right-handed α-helix, a right-handed π-helix, or a right-handed 2/7 helix can be formed. However, a β-sheet structure, as well as any type of left-handed helix, can never be generated in this manner. The configuration of the peptide bond as either R or L is merely a prerequisite for the formation of a right-handed or left-handed spiral structure, respectively. A sufficient condition for the existence of a helical structure is the presence of a system of hydrogen bonds between the atoms of peptide groups from adjacent helical strands, which stabilizes the secondary structure of the protein. Additionally, the sequence must include 3–4 or more consecutive amino acid residues, each encoded by a G/C(U) triplet in the third position of the codon. The properties of the amino acid chains that make up the protein also play a crucial role in the formation of the helical structure. The most comprehensive exploration of the issue regarding the sufficient condition for right-handed alpha helices and 3/10 spirals is presented in the work by V.A. Karasev [146]: In this work, the author utilizes models of connectivity/disconnectivity graphs to differentiate amino acid residues based on the characteristics of their side chains [147].

The relationship implemented in the "Pikotechnology" program ultimately produced a matrix (see Table 6.22) (the first and fourth variants are combined into one). From Table 6.22 it follows that alpha-structures are formed by codons whose third nucleotide is C (cytosine) or G (guanine). Beta-sheets are formed by codons whose third nucleotide is always A (adenine), where the adenine must be preceded by G, T, or C (except the triplets TTA and AGA); and beta-turns are formed by codons whose third nucleotide is always T (thymine), and if the third nucleotide of the triplet is A, it must not be preceded by the base A. The triplets TTA and AGA also form beta-turns. Thus the third nucleotide of a triplet carries information about the formation of the secondary structure of the protein chain.

Table 6.22. Placement of all 64 triplets using the symmetric Gray code. Dark cells mark the triplets coding for alpha-helices; medium-gray cells mark beta-turns and light-gray cells mark beta-sheets. Blank (unshaded) cells correspond to termination codons.

	1	2	3	4	5	6	7	8
1	AAA 000000	AGA 000100	AGT 000110	AAT 000010	AAC 000011	AGC 000111	AGG 000101	AAG 000001
2	ATA 001000	ACA 001100	ACT 001110	ATT 001010	ATC 001011	ACC 001111	ACG 001101	ATG 001001
3	GTA 011000	GCA 011100	GCT 011110	GTT 011010	GTC 011011	GCC 011111	GCG 011101	GTG 011001
4	GAA 010000	GGA 010100	GGT 010110	GAT 010010	GAC 010011	GGC 010111	GGG 010101	GAG 010001
5	CAA 110000	CGA 110100	CGT 110110	CAT 110010	CAC 110011	CGC 110111	CGG 110101	CAG 110001
6	CTA 111000	CCA 111100	CCT 111110	CTT 111010	CTC 111011	CCC 111111	CCG 111101	CTG 111001
7	TTA 101000	TCA 101100	TCT 101110	TTT 101010	TTC 101011	TCC 101111	TCG 101101	TTG 101001
8	TAA 100000	TGA 100100	TGT 100110	TAT 100010	TAC 100011	TGC 100111	TGG 100101	TAG 100001

As an example of overlapping genes, we will consider the viral MS2 messenger RNA (mRNA). Based on its primary structure, we will use a Bayesian recognition procedure to construct the secondary structure of the corresponding proteins. To date, this is one of the largest secondary protein structures in terms of the number of nucleotides.

The MS2 phage has a spherical shape, with a diameter of 25 nm and a molecular mass of 3.6×10^6 daltons. The phage is composed of 180 subunits of a protein coat, each with a molecular mass of 14,700 daltons; one molecule of a so-called "A-protein" with a molecular mass of 38,000 daltons; and one molecule of RNA with a molecular mass of 10^6 daltons. The chain begins with a G nucleotide, which has a triphosphate group attached to its 5'-hydroxyl group. Following this is a long sequence of non-coding nucleotides. The total length of the 5'-circular nucleotide sequence is 129 residues. It contains the triplets AUG and GUG, which, however, do not serve as initiation codons. The first initiation codon, GUG, begins the coding sequence for the A-protein (A-cistron). The A-cistron has a length of 1179 nucleotides and terminates with the stop codon UAG. Then comes a non-encoded insertion section consisting of 26 nucleotides. The next coding sequence begins with "AUG" and has a length of 390 nucleotides; it corresponds to the cistron encoding the coat protein (the C-cistron). It terminates with the stop codon UAA, and immediately following that is the second stop codon, UAG. A sequence of 36 nucleotides separates the C-cistron from the S-cistron that encodes the RNA polymerase subunit. The S-cistron begins with the sequence AUG, has a length of 1635 nucleotides, and ends with the sequence UAG. One nucleotide beyond it (i.e., out of frame) there is another termination triplet, UGA. The 3' non-coding sequence has a total length of 174 nucleotides and ends with adenosine bearing a free cis-glycol group. The complete primary structure of the RNA from the MS2 phage was determined by W. Fiers and his employees from 1971 to 1976. [148]. The MS2 RNA also codes for a fourth protein, which is called the lysin protein, or L-protein. This protein is encoded by a segment of RNA that begins at the end of the C-cistron. This RNA segment includes only a 36-nucleotide insertion located between the C-cistron and the S-cistron, and it ends within the S-cistron itself. The reading frame of this overlapping L-cistron overlaps with that of the S-cistron.

It has been shifted to the right by one residue (+1 shift), such that no part of it is translated during the synthesis of either the C-protein or the S-protein. The L-cistron contains its own initiation codon (AUG), which is not in phase with the codon of the C-cistron; accordingly, it also has its own termination codon (UAA), which is not in phase with the codon of the S-cistron.

In the work [20], the secondary structure of the MS2 matrix RNA is also analyzed, and the linear sequence of this mRNA is provided. It would be interesting to compare the results of the analysis conducted in this paper with those presented in the work [20]. We will not use the entire sequence of the mRNA; instead, we will focus on a segment that includes two proteins: L-protein and S-protein. The amino acid sequence of the L-protein (which begins at nucleotide position 1678) is shown below, while the sequence of the S-protein (beginning at nucleotide position 1761) is shown above the mRNA sequence. In this representation, the nucleotides are denoted by the lowercase letters a, c, g, t, rather than the standard symbols A, C, G, T. This notation has been used frequently in modern scientific literature. The selected segment of the MS2 mRNA chain, which consists of 141 bases (from positions 1761 to 1902), will be sufficient for analysis and comparison of the results with those obtained by N.N. Kozlov in work [20].

We use the nucleotide sequence of the selected mRNA region (MS2) along with a Bayesian recognition procedure. The nucleotide sequence is fed as input to the system, and as output we obtain the secondary structure of the protein (either an alpha-helix, a beta-sheet, or a beta-turn). Consequently, the primary structure of the protein and its corresponding secondary structure will be as follows:

1761

MetSerLysThrThrLysLysPheAsnSerLeuCysIleAspLeuProArgAspLeuSerLeuGluIleTyrGln

atgtcgaagacaacaaagaagttcaactctttatgtat**tga**tcttcctcgcgatctttctctcgaaatttaccaa

CysArgArgGlnGlnArgSerSerThrLeuTyrValLeuIlePheLeuAlaIlePheLeuSerLysPheThrAsn

α α α β_c β_c α α α α β_i β_i β_i β_i β_i β_i β_i α β_i β_i β_i α β_i β_i α β_i

SerIleAlaSerValAlaThrGlySerGlyAspProHisSerAspAspPheThrAlaIleAlaTyrLeu

tcaattgcttctgtcgctactggaagcgg**tga**tccgcacag**tga**cgactttacagcaattgcttact**taa**

GlnLeuLeuLeuSerLeuLeuGluAlaValIleArgThrValThrThrLeuGlnGlnLeuLeuThr 1902

β_c β_i β_i β_i α β_i β_i β_i α β_i β_i α α β_i α α β_i β_c β_c β_i β_i α β_i

where α denotes an alpha-helix, βc and β_i denote beta-sheets and beta-turns, respectively.

Within the reading frame that has been shifted one nucleotide to the left (a shift of –1), protein synthesis does not occur. This reading frame contains termination codons, which are indicated in the nucleotide sequence by bold letters.

The analysis of the secondary structures, compared with the work in [20], showed that the correspondence of β-structures amounted to nearly 75%.

For prokaryotic mRNAs, sometimes a single polynucleotide chain contains coding sequences for multiple proteins, just as in the case of the MS2 mRNA structure. Such mRNAs are referred to as poly-cistronic mRNAs (the term “cistron” was introduced by S. Benzer as an equivalent of the gene). Example: [148].

The type of the secondary structure can be determined by the surrounding amino acids $x_1, x_2, \ldots, x_n$ located to the left and right of the amino acid under investigation. It is possible to recognize pairs and triples of states for two or three current amino acids, as suggested in the work [133]. The Bayes formula for a pair of current amino acids is as follows:

$$P(As, s+1 \mid x1, x2, \ldots, xn) = \frac{P(x1,x2,\ldots,xn \mid As,s+1) \cdot P(As,s+1)}{P(x1,x2,\ldots,xn)}. \tag{6.38}$$

Here, A_s,s+1 denotes the state of the pair of amino acids x_s, x_s+1. The number of different classes of A_s,s+1 is 3600; this is because there are 400 different amino acid pairs in total, and each amino acid pair can have up to 9 possible secondary structures. Numerical calculations have shown that Bayesian recognition procedures applied to first-order Markov chains successfully predict the secondary structure of proteins.

Let’s consider the section of the chain that consists of three elements:

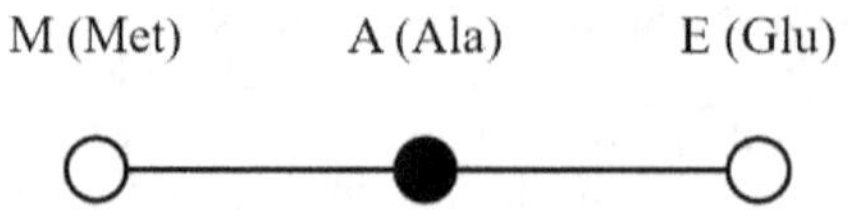

In the process of identifying the secondary structure of the current amino acid A (the dark circle), the probabilities of state pairs for the amino acid pairs MA and AE in the sequence of adjacent amino acids MAE were calculated using the formula (6.38).

So, let's determine the state of amino acid A from the triple set MAE based on the probability estimates for the various state combinations of the amino acids MA and AE. Let's denote this as:

$\hat{R}$MAs (ma) represent the probability estimates for various states of amino acid MA pairs, calculated using formula 6.38. From now on, symbols with a "hat" will indicate these probability estimates.

The resulting expression is: $\hat{R}MA(A) = \hat{R}MA(\alpha A) + \hat{R}MA(\beta cA) + \hat{R}MA(\beta iA$.

Similarly, for the pair of amino acids AE, we will get:

$$\hat{R}AE(A) = \hat{R}AE(A\alpha) + \hat{R}AE(A\beta c) + \hat{R}AE(A\beta i).$$

We calculate: $P(\alpha) = \hat{R}MA(\alpha) + \hat{R}AE(\alpha) = 1 - \hat{R}MA(\alpha) - \hat{R}AE(\alpha),$

$$P(\beta)= \hat{R}MA(\beta c)+\hat{R}AE(\beta c)= 1 - \hat{R}MA(\beta c) - \hat{R}AE(\beta c),$$

$$P(\beta_i)= \hat{R}MA(\beta_i)+\hat{R}AE(\beta_i)= 1 - \hat{R}MA(\beta_i) - \hat{R}AE(\beta_i),$$

$$\hat{R}MA(\alpha) = \hat{R}MA(\alpha\beta_c) + \hat{R}MA(\beta_c\beta_c) + \hat{R}MA(\beta_i\beta c) + \hat{R}MA(\alpha\beta_i) + \hat{R}MA(\beta_c\beta i) + \hat{R}MA(\beta_i\beta i) + \hat{R}MA(\beta_i\beta i) + P(\beta_c\alpha) + P(\beta_i\alpha).$$

Similarly, the following sets of tokens are also defined: $\hat{R}AE(\alpha)$, $\hat{R}MA(\beta c$, $\hat{R}AE(\beta c$, $\hat{R}MA(\beta c$, $\hat{R}AE(\beta_i$.

The strategy proposed in document [149] is quite cumbersome; the probability estimates for each structure are calculated 6 times. Therefore, let's consider a simpler approach to selecting an amino acid: we will use the probability estimates of adjacent state pairs, as well as the numerical calculations related to the estimation of their negation probabilities.

From the results of the evaluations using training datasets, we know that pairs of adjacent amino acids form secondary structures such as α, β_s, and β_i with probabilities that add up to one.

For example: $\hat{R}MA(\alpha A)=0.477$; $\hat{R}MA(\beta cA)=0.3215$; $\hat{R}MA(\beta iA)=0.2015$.

Therefore: $\hat{R}MA\ (A) = \hat{R}MA(\alpha A)+\hat{R}MA(\beta cA) + \hat{R}MA(\beta iA) = 0.477 + 0.3215 + 0.2015 = 1;$

$$\hat{R}AE(A\alpha) = 0.442;\ \hat{R}AE(A\beta c) = 0.2715;\ \hat{R}AE(A\beta i) = 0.2865.$$

Then: $\hat{R}AE(A) = \hat{R}AE(A\alpha) + \hat{R}AE(A\beta c) + \hat{R}AE(A\beta i) = 0.442 + 0.2715 + 0.2865 = 1.$

During the process of predicting the secondary structure of the current amino acid, we will obtain the following probability estimates:

$$\hat{R}A(\alpha)=\hat{R}MA(\alpha) + \hat{R}AE(\alpha)=0.477 + 0.442= 0.919.$$

Similarly: $\hat{R}A(\beta c = \hat{R}MA(\beta c) + \hat{R}AE(\beta c) = 0.3215 + 0.2715 = 0.593;$

$$\hat{R}A(\beta_i)= \hat{R}MA(\beta_i) + \hat{R}AE(\beta_i) = 0.2015 + 0,\ 2865 = 0,\ 488.$$

Each of the structures in the pairs appears twice, so…

$\hat{R}A(\alpha) + \hat{R}A(\beta c) + \hat{R}A(\beta_i) = 0.919 + 0.593 + 0.488 = 2.$

The probability estimates for each structure can be divided by 2.

The result will be: $\hat{R}A(\alpha) = 0.4595$; $\hat{R}A(\beta c) = 0.2965$; $\hat{R}A(\beta_i) = 0.244$.

The state for the current amino acid A(Ala) is defined by the following formula:

$$A = \text{argmax}\{P(\alpha), P(\beta_c), P(\beta_i)\}.$$

Thus, the state of the individual amino acid Ala corresponds to the α-helix.

The best results, as noted by the authors [133], were obtained when recognizing triples of states for three current amino acids.

For a first-order Markov chain model, the probabilities of the sequences x_1, x_2, …, x_n are determined by the following relationship:

$$P(x_1, x_2, \ldots, x_n|A_{s,s+1}) = P(x_1|A_{s,\ s+1})\cdot P(x_2|x_1, A_{s,\ s+1})\cdot \ \ldots\ldots \cdot P(x_n|x_{n-1}, A_{s,\ s+1}),$$

where $P(x_k|k-1, A_{s,\ s+1})$ and k=2,…, n represent the transition probabilities.

In the numerical calculations, transition probability estimates are used, which are represented as frequencies.

$$P(xk = j \mid xk - 1 = i, As, s + 1) = \frac{n(xk - 1 = i, xk = j, As, s + 1)}{n(xk - 1 = i, As, s + 1)},$$

Where $n(x_{k-1} = i, x_k = j, A_{s,\ s+1})$ represents the number of amino acid pairs (ij), such that in the given state of the pair A_s, s+1, the amino acid i is located at position (k – 1) and the amino acid j is located at position k.

$n(x_{k-1} = i, A_{s,\ s+1})$ – The number of sequences in which the amino acid i appears at position (k – 1) under the condition that A_s, s + 1.

As we have already mentioned, the cyclic 4-residue fragment is formed through hydrogen bonds (HN–C=O) that exist between the i-th and (i–1)-th, as well as the i-th and (i–3)-th amino acids. The formation of this fragment requires at least four amino acid residues in the protein, starting from the N-terminus. In subsequent studies related to predicting the secondary structure of proteins, we will use a graph that consists of five residues.

The method being studied belongs to the category of machine learning techniques and operates on training datasets. Open databases containing experimentally determined secondary protein structures are used as these training datasets.

One of the main challenges in analyzing nucleotide sequences is that the frequencies of adjacent nucleotides are not independent of each other. In particular, the frequencies of pairs of adjacent bases usually differ from the products of the frequencies of the individual bases themselves. The constraints imposed on adjacent bases have the effect of making even more distant bases become interdependent as well. A Markov chain of order k assumes that the occurrence of a base at a given position in the sequence depends only on the bases located in the preceding k positions. A chain of order 1 assumes that the probability of a base occurring at position i depends solely on the probability of one of the four bases appearing at position i–1. A sequence consisting of independent bases would correspond to a Markov chain of order 0. By conducting studies, we can predict the frequencies of certain regions of interest and estimate the expected number of fragments obtained upon DNA fragmentation. If a sequence can be regarded as a Markov chain of order k, then the nth base depends only on the preceding k bases; the probability of transitioning to a particular state in the sequence can thus be determined. For example, given the sequence GATATTAT, the probability that T follows TAT is …

$$\hat{P}(T|TAT) = \frac{nTATT}{nTAT} = \frac{1}{2}.$$

For a sequence of length k, the likelihood of some particular order is equal to the product of the probabilities of the first k bases and the probabilities of each of the subsequent groups of k + 1 bases. For example, the likelihood for a sequence of length 2 would be:

$$L = P(GA)\cdot P(T|GA)\cdot P(A|AT)\cdot P(T|TA)\cdot P(T|AT)\cdot P(A|TT)\cdot P(T|TA)\cdot P(C|AT),$$

P(T|GA) denotes the probability that T follows the pair GA. This probability is calculated as the number of GAT triplets divided by the number of pairs of GAs.

The higher the order of the model, the more parameters it takes into account and the better its fit with real data becomes; however, higher-order chains also require estimating more parameters (and thus more data).

Katz (1981) [138] demonstrated the usefulness of introducing a “penalty” when the number of parameters being considered is too large. To this end, he proposed using the Bayesian Information Criterion (BIC).

$$BIC(k) = Const - 2 \ln L(k) + 3\cdot 4^k \cdot \ln n_k$$

where n_k refers to the number of subsequences of length k + 1 that are present in the given sequence. The value of k for which BIC(k) is minimal is taken as the estimate. The lowest BIC score was obtained for the value of k = 1. This confirms the conclusion drawn from the likelihood ratio analysis, which suggests that the order sequence “1” best matches the data. Once the order of the chain is established, it can be used to determine the frequencies of certain sequences occurring in the data. This was done for yeast DNA by Arnold and others (Arnold et al., 1988; reference: [138]). The authors used third-order reaction chains and calculated the probability of obtaining the sequence GCGGCCGC, which is recognized by the enzyme Notl (the enzyme that induces the maximum number of cleavages). This probability is calculated as the ratio of the number of different tetra- and trinucleotides in the given sequence.

As has already been mentioned, many biological processes involve specific interactions between DNA-binding proteins and DNA sequences. For a protein to bind to DNA, its surface must be sufficiently complementary to the double-stranded DNA structure over a large area. In this way, the protein’s structural features can fit neatly into the DNA groove, allowing its binding domains to precisely recognize the target DNA sequence. The DNA-binding α-helices in a protein dimer are antiparallel to each other, and the distance between them is roughly equal to the helical pitch of DNA; as a result, the dimer binds to one side of the DNA double helix. However, the different angles at which these α-helices are tilted relative to the axis connecting their centers cause the proteins to bend the DNA in various ways during binding. DNA-binding proteins can belong to different structural classes (such as α-helix or α-helix-β-sheet complexes), and the actual binding mechanism can involve both α- and β-stranded regions of the protein. The "common rules" that govern how proteins selectively recognize specific DNA sequences remain largely unknown. By examining the details of each identified DNA-binding interaction, it is possible to determine which hydrogen bonds between the protein's side chains and nucleotides, as well as other types of interactions, contribute to the formation of the DNA-protein complex at that particular site [125].

In the work [64], it is noted that when analyzing experimentally determined protein-DNA structures, a number of patterns can be identified: preferentially, alanine-thymine pairs are formed (due to the interaction of methyl groups), as well as hydrogen bond pairs between arginine and guanine, and asparagine and adenine.

The contact region between proteins and DNA is enriched with polar amino acids; purines show a higher selectivity for certain amino acids than pyrimidines, while aromatic amino acids may have various preferences. The study indicates that in many cases, a protein's preference for a particular nucleotide at a specific position depends on the presence of certain residues in other positions within the protein. This requires the selection of an optimal model complexity. The paper discusses the methods used to address this issue. The probabilities of rare residues turn out to be statistically insignificant. To avoid this effect, residues that occur in the sample at a frequency lower than a certain threshold were grouped together and considered as a single type of residue. Here, $r = 1/m_j$, where m_j represents the number of residue groups at position j after grouping. In the task of classifying "AP-site" pairs, considering the equal probability of all classes, Bayes' formula takes the following form:

$$P(c \mid AAS, NS) = \frac{0.5 \cdot P(AAS,NS|c)}{\sum 0.5 \cdot P(AAS,NS|ci)}, \quad (6.39)$$

AAS — amino acid sequence;

NS: The nucleotide sequence of the site.

"c" refers to a class.

It is further assumed that the pairs of positions in the amino acid sequence and at the site are conditionally independent as a whole.

$$P(AAS, NS \mid c) = \prod_{i,j} P(a_i, n_j \mid c),$$

where a_i – represents the amino acid residue at position i;

n_j is a nucleotide located at position j.

The conditional probabilities for various position pairs were calculated using pseudo-counting methods and by grouping together rare amino acid residues.

$$P(ai, nj \mid c) = \frac{Wij(ai,nj|c) + k\sqrt{Wp}}{W + k\sqrt{W}}, \quad (6.40)$$

where a_i to represent the type of amino acid residue;

n_j – is a type of nucleotide.

$W_{ij}(a_i, n_j \mid c)$ represents the sum of the weights of all pairs in class c that contain a_i at position i of the amino acid sequence, and the nucleotide n_j at position j of the site.

r = 1 / m; where m represents the number of possible amino acid residues.

"k" – a coefficient that regulates the contribution of pseudo-counts.

In [64], two methods are proposed that allow for a quick assessment of the predictive power of a given position within the AAS (amino acid sequence) for predicting the nucleotide at a specific position in the site. The first method is M_1; its advantage lies in the speed of its computation, which makes it suitable for use in feature selection processes. This method helps to eliminate irrelevant features, but it does not take into account the dependencies between those features.

Another method for feature selection involves iterating through subsets of features, training the algorithm on a portion of the training dataset, and evaluating the error rate on the remaining part of the dataset. The subset that results in the lowest error rate is selected. In this study, a so-called "greedy" algorithm was used. The algorithm chosen would identify the feature whose addition led to the best performance of the classifier. The greedy algorithm can be improved in several ways; for example, by sequentially adding or removing features, or by maintaining multiple candidate features (i.e., keeping several top-performing features).

Experiments using the Bayesian classifier (which learns the fastest) have shown that improved algorithms select the same sets of features as the simple, "greedy" algorithm does.

6.7. EVALUATION OF THE RELIABILITY OF RECOGNITION AND CLASSIFICATION PROCEDURES. THE BAYESIAN APPROACH

Methods for solving complex problems can be divided into exact and approximate methods. All exact methods, due to the NP-complexity of the problem, are essentially variations of the brute-force search approach; however, various heuristics can be used to reduce the number of options that need to be considered. The most well-known such method is Branch and Bound (BB), denoted as [150].

This method breaks down the original problem into a series of sub-problems, so any valid solution to the original problem is also a solution to one of the sub-problems.

For each subtask, the value of the objective function is evaluated on the best of its solutions. If this evaluation score is higher than the minimum functionality score of any previously examined solution for the original task, then that subtask is not solved. All other subtasks are solved in the same manner. The quality of such methods depends, first, on how each emerging subtask is divided into smaller subtasks; second, on the choice of a good evaluation function to identify obviously suboptimal solutions. However, there is no guarantee that finding the optimal solution will not require examining all possible options. There exists another type of exact method with polynomial complexity, but it is intended for a limited class of algorithm graphs and/or computational system graphs [151].

In any case, applying precise methods to solve problems involving large graphs is problematic either due to the high complexity of the graphs or because the range of allowable graph structures is limited. As a result, approximate methods have become widely used; these methods can also be divided into specialized and general-purpose approaches.

The most common methods are the universal ones, and some of them are modifications of more specific ("precise") methods.

First and foremost, in most cases, the question arises as to whether a reasonable method can be found for identifying "equally likely events." For instance, given the symmetrical nature of the DNA chain's base structure, it is difficult to determine the probability of certain nucleotides occurring. Previously, another approach for approximating the unknown probability of a random event was mentioned in the theory of probabilities. Extensive observations of the occurrence of event A across a large number of independent experiments, all conducted under the same set of conditions, have shown in some cases that the frequency of event A follows stable patterns. Specifically, if we denote the number of occurrences of event A as μ and the total number of experiments as ν, then it is observed that the ratio…

$\frac{\mu}{\nu}$ — The frequency of event A (denoted by "ν") is sufficiently high for such series to occur.

The observed value remains almost constant. In cases where the classical definition of probability applies, this fluctuation in frequency occurs around the probability of event A, denoted P.

If, as a result of numerous observations, the frequency of event A behaves in a fairly consistent manner and almost always fluctuates around some constant value, then one can say that event A has a certain probability. The numerical value of this probability can be taken to be the frequency of event A (when a large number of independent experiments are conducted). However, experiments allow us to draw conclusions of a different nature as well. Certain circumstances provide grounds to believe that the probability of event A equals P. Suppose that in several series of independent experiments, it is observed that the frequencies in the vast majority of cases deviate significantly from the value R. This circumstance gives us grounds to question the correctness of our initial assumptions and to conduct a more thorough examination of the premises on which we based those conclusions. We should consider the concept of probability as formulated by the German mathematician R. Mises (1883–1953), which is particularly prevalent among natural scientists. Mises — the founder of the frequency theory of probability — did not regard probability theory as a mathematical discipline, but rather as a science that makes extensive use of mathematical methods. According to Mises, if the frequency of an event deviates less and less from the value R as the number of experiments increases, then in the limit…

$$P = lim_{n \to \infty} \frac{\mu}{\nu}.$$

This equation proposed by Mises should be regarded as a definition of the concept of probability. Any a priori definition is doomed to fail; only the empirical definition offered here can satisfy the needs of natural science, mathematics, and philosophy. Moreover, since the classical definition has very limited applicability, while the statistical definition applies to all cases of scientific interest, Mises suggested that the classical definition should be discarded. Mises's theory is elaborated in detail in his book Probability and

Statistics. His theory has been subject to critical criticism, as discussed in the article by A.Y. Khinchin [152].

In probability theory, attempts are made to avoid defining the probability of a mass event as the limit of a frequency ratio (μ/ν) as the number of trials $n \to \infty$.

The numerical value representing the probability is chosen to be a number near the frequency at which the event occurs, based on a large but finite number of observations [153].

There are currently many recognition and classification procedures that are successfully applied to solve various practical problems. The problem of classification can be viewed as a problem of "statistical decision functions." There are several different hypotheses, each corresponding to its own probability distribution for the observed data. When designing a classification or recognition procedure, it is desirable to minimize the likelihood of incorrect classifications; in other words, we want to ensure that incorrect outcomes occur as rarely as possible on average. However, it is important not only to obtain an algorithm that achieves the desired class separation but also to have an estimate of the reliability of the solution to the problem at hand – that is, to know how often the algorithm will make mistakes when classifying new objects. Clearly, this evaluation directly determines the effectiveness of the solution in addressing the problem.

The set of all possible random implementations, denoted by $x_1, \ldots, x_n$, is called a sample, and the number n represents the size of the sample.

The authors define the concept of a "small sample" in different ways. Some consider a sample to be small if its size does not exceed 200; others define it as a sample with a size not exceeding 30, 50, or several dozen units, or less than a certain calculated value determined using a special nomogram for large numbers. A sample is considered small if the desired level of accuracy and reliability cannot be achieved by using methods based on data grouping and approximation techniques. Therefore, the concept of a small sample is relative and depends on the specific task at hand [135].

When restoring the probability density from a sample, there are two approaches: parametric and non-parametric.

The parametric approach is based on the hypothesis that the desired probability density belongs to a class of probability densities determined by a set of parameters. Within the parametric framework, the task of reconstructing the probability density boils down to identifying the values of those unknown parameters from which the probability density depends.

Non-parametric methods for restoring the distribution laws of observed random variables do not require such a strict fixation of the class of distribution laws using a finite number of parameters. Instead, non-parametric estimation techniques, as described in the works of Rosenblatt and Parzen [135], are typically used.

For the solution of many non-parametric problems, it has for a long time been unsuccessful to apply Bayesian approaches to statistical inference. This is mainly due to the difficulty in finding a viable prior distribution on a parametric space; in non-parametric problems, the prior distribution is chosen as a set of probability distributions on the sample space. One of the first successful attempts in the field of non-parametric Bayesian estimation was the work by Ferguson [154]. In this paper, Ferguson first formulated the requirements for such a prior distribution:

1. The carrier of the prior distribution must be large relative to some appropriate topology of the probability distribution space, as defined on the sample space.
2. The posterior distribution, given a specific sample of observations from the true probability distribution, should, whenever possible, have a simple analytical form.

These properties are antagonistic in the sense that one of them can be obtained at the expense of the other. In paper [154], a class of prior distributions called Dirichlet processes is proposed; under certain broad assumptions, these processes possess the first property and also satisfy the second property. The choice of Dirichlet processes is motivated by the fact that the posterior distribution of a random probability measure is itself a Dirichlet process. Another argument in favor of using the Dirichlet distribution in practical applications is that it can well approximate many parametric families of distributions. This issue was specifically investigated by Dalal in [155] and Hall in [156]. Recall that the Dirichlet principle, in its general form for numbers, states: "If the sum of n numbers is greater than S, then at least one of these numbers is greater than S/n."

In other words, this can be phrased as: "If the arithmetic mean of several numbers is greater than a, then at least one of the numbers must be greater than a." In parametric Bayesian theory, the non-parametric Bayesian estimation based on Dirichlet processes is used as a counterpart to the core concept of sampling likelihood for estimating parameters of a multinomial distribution [157].

In mathematical statistics, the approximate values of probability density parameters or the probability density itself, obtained based on a sample, are commonly referred to as estimates. Estimates that are closest to the true values are called efficient estimates. Of course, it is always desirable for an estimate to be efficient. Unfortunately, efficient estimates are often unstable; in cases where the underlying assumptions are violated, robust estimates (which have maximum efficiency under such conditions) should be preferred.

In mathematical statistics, there are two methods for estimating the characteristics of probability distributions based on random observations: the point estimate and the interval estimate [158].

The "pointwise" method involves calculating approximate values for these parameters based on available observations; this is often referred to as the probabilistic approach [7]. The second method consists of constructing regions (sets) in the parameter space that with a certain probability contain the true value of the parameter. The probability of such containment is called the confidence level; the resulting set of possible values is known as the confidence interval, and the entire approach is collectively referred to as confidence estimation [159]. This method is also sometimes called the "guaranteed" approach [7] or the interval-based approach [158].

The formulation of a Bayesian estimation task in general includes four elements:

a) The distribution of probabilities for the main random variable;

b) A priori probability distribution;

(c) The loss function, which represents the losses associated with replacing the true value of the reliability indicator R with its estimated value $\hat{R}$;

d) The test plan P, prescribing the method for obtaining experimental data.

The task is to combine prior information with experimental data in order to obtain estimates for the reliability metric R [51].

Point estimates for unknown parameters are obtained through different methods, so they are not directly related to each other. However, if both the point estimate and the confidence estimate are derived within the same framework (for example, a sign-based approach), then the confidence intervals are constructed around the point estimate. The size and shape of these confidence intervals provide information about the accuracy of the point estimate: the narrower the confidence interval, the higher the accuracy of the estimate. This accuracy depends primarily on the properties of the statistical errors, but also on the estimation method chosen. Asymptotic confidence intervals (where the sample size approaches infinity) can be relatively easily constructed for parameters in a linear model; all that is required is an asymptotically normal estimate. A wide class of such estimates includes Huber's M-estimates, denoted as [159]. To determine which estimation method yields the best results, it is necessary to compare the estimated values as random variables. The more concentrated the distribution of the estimated values around the true value, the more accurate the estimate will be. Unfortunately, estimates from different methods are usually not comparable with each other. Therefore, it is not surprising that, with a finite number of observations, there is no single "best" method for estimation.

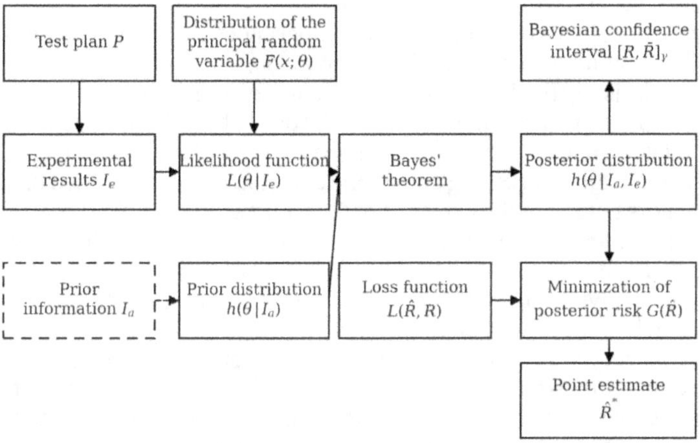

Figure 6.4. A general overview of the process for obtaining Bayesian estimates

In the asymptotic approach, it is still possible to compare the methods used for estimation.

In a number of evaluation methods, it has been found that the estimates of parameters become asymptotically normal as the number of observations increases (more precisely, as the amount of information grows). This is because, in addition to the number of observations, many other factors depend on the experimental design. To simplify the analysis, it is assumed that the experimental designs are optimal. Under these assumptions, the number of observations (n) becomes the only dominant variable.

In applications, the confidence level and risk level are often expressed as percentages, and the boundaries are chosen such that the confidence level corresponds to one of the standard values: 0.9; 0.95; 0.99; 0.999. For example, if rdov = 0.95, it indicates a 95% confidence level and a 5% risk level. Interval estimates are more informative than point estimates, but they require more computational resources to generate.

The assessment of the absolute justification of knowledge primarily relates to knowledge that is acquired in a dynamic process, has not yet taken shape, and therefore requires some reliable basis for support. The assessment from a rational perspective, on the other hand, essentially involves evaluating knowledge in a static context—considering it as something that has already been established, stabilized, and completed to a certain extent.

The choice of a specific option and the procedure for evaluating the degree of compliance depend on many factors, among which the most important ones include:

- The presence or absence of the possibility to experimentally verify the research results…
- The presence or absence of the possibility to directly compare evaluations (empirical and theoretical).
- The presence or absence of the possibility to conduct a quantitative comparison of evaluations.
- Potential accuracy of quantitative estimates;
- A type of qualitative scale used for comparing evaluations when it is not possible to conduct a quantitative comparison.

Sometimes it is difficult to measure a particular characteristic directly; therefore, an indirect measurement is carried out, that is, the measurement of parameters related functionally to the parameters being evaluated. Often, the tasks of mathematically processing and interpreting uncertain data based on these indirect indicators belong to the category of "inaccurate" problems. It is essential for any experimenter or specialist dealing with uncertain data to know the relevant methods and be able to use them to solve such problems. Most tasks involving the analysis of measurement results from objects cannot be accomplished without the use of computers. When verifying the validity of a theory, the concept of a "decisive experiment" is employed. Similarly to a decisive experiment in experimental sciences, there is the concept of a "defining factor" in non-experimental sciences—a factor that corresponds to a randomly occurring set of system operating conditions. One of the most effective ways to utilize non-experimental data is to divide them into two subsets: a training set and a control set.

The set of variables through which the dependent variable is expressed must be…

- unique; complete; compact (small in number); and observable.

The other variables that affect the dependent variable must follow a certain probability distribution, which is determined by several unknown parameters.

All unknown parameters should be constants.

The market for scientific products is overwhelmed with theories, each claiming to explain the same phenomenon in its own unique way. There are studies that contradict one another regarding the theoretical foundations of their evaluations, predictions, and explanations. To date, there does not exist a universally effective method for providing a scientific basis for empirical data in all cases.

However, there is a general, widely tested approach for solving these problems, which is denoted as [7].

When identifying patterns, quantitative relationships, and the forms of connections between variables that characterize a system, the researcher undertakes six specific types of tasks:

1. Hypothesis testing – the process of selecting the "true" model.
2. Interpretation of multi-dimensional observations.
3. Simplification: Building a "useful" model.
4. Refining an approximating variable involves finding a suitable quantitative characterization.
5. Data selection – the process of creating a set of empirical data.
6. Modify the original model after analyzing the data (improve the empirical model that has been constructed, or possibly also the theoretical model).

According to the standards recognized by the international scientific community when evaluating the credibility of scientific findings, two main aspects are taken into consideration:

- The accuracy of predicting outcomes in both practical and scientific activities is crucial.
- This refers to the level of confirmation or validation of a scientific finding.

The accuracy of predicting outcomes in practical and scientific activities can be assessed using various methods.

The most common approaches are probably the probabilistic one and the guaranteed one, which is represented by [160].

In a probabilistic approach, "point estimates" are used. The mathematical expectations of errors are considered to be equal to zero, and the covariance matrix of errors is defined with an arbitrary positive multiplier.

In a guarantee-based approach, the values used to represent accuracy correspond to the worst possible values of the elements in the mathematical expectation vector and the covariance error matrix, as well as the maximum possible deviations of these elements from their average values. If the predictions of the studied processes based on the scientific results are accurate, then the degree of correlation between the predicted outcomes and the actual results is high; otherwise, the degree of correlation is low. A more detailed description of the accuracy of scientific results can be found in the document [7].

When restoring the values of the parameters being studied, researchers usually have additional (a priori) information about those parameters, in addition to the information contained within the sample. Previously, parameter estimates were based solely on the sample data and did not take into account this additional information. By incorporating this extra information, the reliability and accuracy of the estimates can be improved.

Currently, three main methods are used to take into account additional information, depending on the nature of that information: the Bayesian method, the minimax method, and the generalized maximum likelihood method [158].

The application of a Bayesian approach assumes that the parameter f being sought is treated as a random variable. The prior information is then specified as a prior distribution papr(f) for the random variable f. The Bayes method, in many cases, allows for the successful incorporation of prior information and enables the derivation of reasonable estimates. Let p(x_1, …, x_n, f) denote the joint probability density of the set of random variables x_1, …, x_n, f.

P(x_1,…, x_n, f) = p(x_1, …, x_n | f) papr(f) = p(f | x_1,…, x_n) p(x_1,…, x_j,…, x_n),

where p(f | x_1,…, x_n) represents the conditional probability density of the random variable f given the fixed values of x_1,…, x_n;

p(x_1,…x_j,…x_n) represents the marginal joint probability density of the set of random variables x_1,…, x_n. This can be calculated using the following formula:

$$p(x_1, \dots, x_j, \dots, x_n) = \sum_{i=1}^{n} \prod_{j=1}^{k} p(x_1, \dots, x_n \,|\, f)\, p_{apr}(f), \tag{6.41}$$

referred to as the "full probability," where k is the number of features.

From the above information, the formula can be derived as follows:

$$p(f \mid x1, \dots, xn) = \frac{p(x1,\dots,xn|f) \cdot papr(f)}{\sum_{i=1} \prod_{j=1} p(x1,\dots,xn|f) \cdot papr(f)}. \tag{6.42}$$

Since there is one specific implementation of the set of random variables {x_1, …, x_n} provided by the sample, we can finally derive Bayes' formula: P(f | X_1, …, X_n) = L_n(X_1, …, X_n) × papr(f) / C.

The expression L_n represents a likelihood function.

"C" is a normalization factor.

$$C = \sum_{i=1}^{n} \prod_{j=1}^{k} L_n(X_1, \dots, X_n)\, p_{apr}(f) \tag{6.43}$$

As the point Bayesian estimate (B-estimate) fB, one of the characteristics of the posterior probability density is chosen, such as the mode, mathematical expectation, or median of the posterior distribution. Clearly, in the general case, all three estimates will be different.

There is also another way to obtain an exact Bayesian estimate, which involves specifying a positive loss function.

W($\|\hat{f} - f\|$), where $\hat{f}$ is an estimate of the parameter f. Based on the selected loss function W(t) and the obtained posterior density p(f; x_1,…, x_n), a functional relationship is constructed.

$$R(f; x_1, \dots, x_n) = \sum_{i=1}^{n} \prod_{j=1}^{k} W(\|\hat{f} - f\|)\, p(f \,|\, x_1, \dots, x_n), \tag{6.44}$$

This is commonly referred to as the posterior risk estimate of parameter f for a given sample. In that case, B is determined as the solution to an extreme value problem (i.e., a problem involving the estimation of parameters under extreme conditions).

$\hat{f}_B = arg\, min_{\hat{f}}\, P(f\,;\, x_1, \dots, x_n) =$, that is, B represents the value that minimizes the posterior risk.

The Bayesian approach is based on the assumption that the parameter of interest (for example, the error rate of a recognition procedure) is distributed according to some prior distribution, which reflects our degree of knowledge about its value. Using this prior distribution and Bayes' formula, we can determine the posterior distribution as a function of the observed data.

The Bayes' formula allows us to determine posterior probabilities by multiplying the prior probabilities by the likelihood values. We consider the prior probabilities to be a measure of our uncertainty. Thus, we adopt what is known as a "subjective" approach in statistical tasks involving the evaluation of [161]. Now, using these obtained posterior probabilities, we need to determine which event from system A actually occurs. In the simplest case, we can choose the event with the highest posterior probability. This evaluation method is called the maximum a posteriori (MAP) estimation. The Bayesian solution is the one that minimizes the average risk as well as the difference between the parameter value and its estimated value. It is represented by a function; in the most common formulation, this function represents the "loss" incurred due to using the estimated value instead of the true parameter value. The evaluation rule is chosen such that it minimizes the mathematical expectation of this loss function.

The most common measure of the accuracy of Bayesian estimates is the mean risk. A Bayesian estimator is a statistical estimate that minimizes the posterior risk. For example, the Bayesian estimate of a scalar parameter θ, given the quadratic loss function $W(\theta, d) = (\theta - d)^2$, is the posterior mean; whereas given the loss function $W(\theta, d) = |\theta - d|$, it is the median of the posterior distribution of θ. The Bayesian estimate of the mean value θ of a normal (θ, σ) distribution, based on a sample of fixed size n, with a normal (μ, τ) prior distribution of θ, is the posterior mean $(nx\sigma^{-2} + \mu\tau^{-2}) / (n\sigma^{-2} + \tau^{-2})$. This holds true regardless of the specific form of the convex loss function $W(|\theta - d|)$ [162].

The rule according to which a decision-making strategy is chosen in order to minimize the average risk is called the Bayes' criterion. The application of the Bayes' criterion is appropriate when a recognition system repeatedly identifies objects or phenomena within an unchanging environment, provided that the classes are described in a stable manner. To calculate the average risk, it is necessary to define a loss function, which in the general case is some function of the estimate and the true value of the parameter being assessed.

As Bayesian estimates, various characteristics of the posterior distribution can be used, which minimize the average risk under a given loss function. The most commonly used characteristics are the mathematical expectation, mode, and median of the posterior probability density.

In most cases when using a Bayesian approach, a quadratic loss function is employed; the loss value proportional to the square of the distance between the estimated parameter value and the true value of the parameter. The advantage of a quadratic loss function is that it "penalizes" large errors significantly. Therefore, for tasks where large estimation errors are highly undesirable (for example, evaluating the quality of a classification algorithm when there are few examples), a quadratic loss function should be used.

The rules mentioned, which are used to obtain estimates, constitute the basis of Bayes' Theorem (Bayes' Principle), and it is one of the most important concepts in mathematical statistics.

From the Bayesian approach, it follows that in order to apply it, one needs to know the distribution of the prior probabilities. However, very often the prior probabilities are unknown and must be determined based on additional information specific to the particular problem at hand. In the absence of such information, it is assumed that the prior events are equally likely. This assumption is known as Laplace's Principle of Indeterminacy, as well as Bayes' Axiom or the Equality of Likelihood Principle (denoted as [135]).

In the context of the Bayesian approach to statistical problems (S.I. Gurov uses the term "Bayes" in his work [163]), a new method for specifying the prior distribution is proposed, based on a principle called the "Consistency Principle." The author notes that other methods for specifying the prior distribution also exist, such as those based on the principle of minimum information entropy, concepts of subjective probability, utility functions, confidence levels, etc.; however, their application is quite limited. In his paper, S.I. Gurov proposes a new principle for determining the prior distribution of an unknown parameter θ, which in turn specifies the distribution $R\theta(\xi)$ of the random variable ξ observed in the statistical experiment. The principle consists of two main steps: (1) identifying a certain class of possible prior distributions G_s that are natural for the given statistical model, each characterized by a finite-parameter γ; and (2) using the results of the statistical experiment to determine the value of γ. This principle is called the Consistency Principle because it requires both the homogeneity of the prior probability densities and the likelihood function of parameter θ, as well as the equality between the classical frequentist estimate of θ and the Bayesian point estimate of θ. The identification of the class G_s of prior distributions makes use of concepts from R. Fisher's fiducial theory. As the author states, the proposed Consistency Principle unifies the classical frequentist, Bayesian, and fiducial approaches of mathematical statistics. The prior distribution obtained using this principle then determines the posterior distribution, which is used to draw statistical conclusions—in particular, for the confident estimation of the unknown parameter θ [163].

The theory of fiducial probabilities is based on the idea of "reversing the probability" proposed by R. Fisher. The search for a logical basis for this heuristic approach continues to this day. Fiducial probabilities are one of the tools used in statistical estimation to estimate an unknown parameter of a distribution of observed sequences of equally distributed random variables. This concept was introduced by R. Fisher in the 1920s, around the same time as Y. Neyman introduced the notion of a "confidence interval." By their very nature, fiducial probabilities have an heuristic, semi-intuitive character; subsequent statistical research has attempted to provide arguments for their justification. A.N. Kolmogorov noted that in situations of complete prior uncertainty and with a small number of observations, fiducial probabilities yield the best estimates. Sh. Zacks showed that in the theory of statistical inference, fiducial probabilities provide the best solution in certain circumstances.

In the 1980s, V.Z. Belen'kyi [165] considered various approaches to establishing an invariant formulation of the optimal choice problem in situations of complete prior uncertainty and independently came up with the concept of "fiducial probabilities." He formulated the principle of inertia and its connection to the theory of fiducial probabilities. It was noted that the principle of inertia can serve as a logical basis for the theory of fiducial probabilities. The principle of inertia for free random variables, as stated in work [165], is of universal applicability; fiducial distributions are uniquely determined by this principle. This is particularly important because, prior to this, the theory of fiducial probabilities did not have a rigorous logical foundation. A brief formulation of the principle of inertia is: "Random variables that are free (i.e., not constrained by any prior assumptions) should remain the same both a priori and a posteriori."

It directly follows from Equation 6.29 (see Section 6.4.2) that the error tends to zero as the training sample size increases; in other words, the reliability of the Bayesian recognition procedure improves with the enlargement of all classes within the training dataset. A key feature of Bayes' Theorem is that its practical application typically requires a tremendous amount of computational effort. Consequently, the flourishing of Bayesian estimation methods coincided with the revolution in computer and network information technologies. Bayes' formula is related to the calculation of probabilistic relationships between multiple variables and the determination of their interconnections. By utilizing this technique along with computational power, it becomes possible to uncover correlations between various pieces of information.

The quality of recognizing the state "f" using the function A(x) in the Boolean case can be expressed as follows:

$$I(A) = \sum_{x=1}^{n} \sum_{f=0}^{1} (A(x) - f)^2 P(x, f), \quad (6.45)$$

The averaging is performed over all 2n vectors $x = (x_1, x_2, \ldots, x_n)$.

When solving recognition problems, it is necessary to construct an optimal classifier function f(x). Specifically, the goal is to design such a function that, when elements x from the set X are fed into the classification process, the equality $f(x) = f^*(x)$ (i.e., correct classification) holds as often as possible in practice. The function f^* is called the true classifier.

The quantitatively assessed degree of confidence v in the correctness of this equality for any arbitrary element $x \in X$ is what we seek to determine. The task of evaluating the reliability of a decision rule consists precisely in determining the value of v.

The most advanced results in the field of algorithmic classification reliability have been obtained within the framework of Vapnik-Chervonenkis theory. This theory establishes the necessary and sufficient conditions for the uniform convergence of the frequencies $v_l(A)$ of event A occurring in experiments conducted using the Bernoulli scheme, on a given subset of events along with their respective probabilities P{A}. In other words, it provides a criterion for verifying the satisfaction of a certain mathematical relationship.

$$P\{ sup_A \left| P(A) - v_l(A) \right| > \varepsilon \} \rightarrow 0 \ (l \rightarrow \infty), \ 0 < \varepsilon < 1. \quad (6.46)$$

To apply the Vapnik-Chervonenkis theory, there is no need to reconstruct the probability distribution densities, which is undoubtedly a significant advantage of the approach. For our task of estimating the reliability of a decision rule, we are forced to assume that the classifier function f(x) is chosen from a fixed set of pre-defined decision rules F. However, making such an assumption does not always justify the resulting recognition solution. This condition holds, for example, in the case of a finite feature space, where the set of decision rules F is always explicitly defined and finite. Even in these situations, where the class F is determined before solving the problem, it is often impossible to compute its cardinality, as doing so "reduces to cumbersome combinatorial calculations that may not be feasible" [135]. There are also numerous classification methods for continuous features (e.g., using potential functions in machine learning implementations or algebraic approaches to construct correct recognition algorithms), where the classifier is constructed directly during problem-solving and the set F is not predetermined. Moreover, it is always possible to first determine an optimal classifier f_{min} in terms of minimum complexity, and then re-formally solve the problem by assuming $F = \{f_{min}\}$ and $|F| = 1$. This raises questions about the applicability of Vapnik-Chervonenkis theory to our specific task. Nevertheless, the successful application of this theory to a wide range of recognition problems suggests that the estimated required lengths l are often significantly overestimated. One reason for this is the implicit assumption that the decision rule used for evaluation is randomly selected from the set F. As a result, estimates for the deviation probability $\nu_l(A)$ (i.e., the probability that event A occurs) relative to its actual probability P(A) rely on Hoeffding's inequality or the somewhat more crude Bernstein estimate, which cannot be significantly improved [135].

The evaluation result follows from Bernstein's inequality:

$$P\{\ |P(A) - \nu_l(A)|\ > \varepsilon\} < exp\{-\varepsilon^2 l\}, \qquad (6.47)$$

Since the variance of an individual component in empirical risk...

$I_e(A) = \nu(A) = \frac{1}{l}\sum_{i=1}^{l}(f_i - F(x_i, A))^2$ should not exceed 0.25.

$$P\{\ sup_A\ |P(A) - \nu(A)|\ > \varepsilon\} < exp\{-\varepsilon^2 l\}. \qquad (6.48)$$

where N is the number of events $A_1, A_2, \ldots, A_n$.

In order to overcome the difficulties associated with the Vapnik-Chervonenkis condition, the problem is considered in the context of logical decision rules for which the computational complexity (i.e., the power F) is finite. For a finite set of decision rules from (6.48), an upper bound on the error rate υ(Q, C) of the Bayesian recognition procedure was obtained in the work [166].

$$\upsilon(Q, C) \leq \alpha\sqrt{\frac{gn}{min(m_0, m_1)} + \frac{2}{m_2}}, \qquad (6.49)$$

where α is the absolute constant.

The lower bound differs from the upper bound by an absolute constant. Therefore, Q_B is optimal, as it performs the best at identifying the objects.

The advantage of an inductive approach over a deductive one in the Bayesian recognition procedure being considered is that the results are based on empirical information contained in all classes of the training dataset. Furthermore, unlike deductive systems, the processing of empirical data in inductive procedures is performed in parallel for all classes of the training dataset. Most importantly, the task is solved in a single computational step, which means that the solutions can be easily obtained on modern computers.

It should be particularly noted that this recognition procedure can be widely applied in various fields of science and technology.

In biology, researchers study macromolecular protein complexes that consist of hundreds or thousands of amino acid sequences. Clearly, it is impossible to examine each of these sequences individually; therefore, the tools of probability theory are widely used in these fields. The Bayesian procedure can be utilized to model the process of crop selection and to predict the properties of protein compounds. Analysis of the human genome has revealed that DNA is encoded using a four-letter nucleotide alphabet, in the form of a stationary ergodic Markov chain. Since the Bayesian procedure works optimally for such chains, it can be used to identify and eliminate unfavorable (or "unsuccessful") nucleotide sequences in proteins.

However, it turns out that Bayesian recognition procedures based on Markov chains are far from omnipotent; in fact, the formation of many even linear macromolecules cannot be described solely using these methods. The reason is that the probability of monomer molecules undergoing reactions to form a polymer chain is influenced by the spatial arrangement of the individual chain segments. This poses difficulties in experimentally determining the transition probabilities between different states, thereby limiting the applicability of Markov chains in predicting the composition and structure of polymers. On the other hand, the different spatial arrangements of the chain segments within a macromolecule naturally affect the physical properties of the polymer. These physical properties can be calculated using other methods, notably matrix analysis techniques.

6.8. MATRIX ANALYSIS

Matrix genetics, through the Fibonacci genomic matrices, reveals new connections between genetic code and the laws of phylogenetic organization (phyllotaxis), offering new perspectives on understanding the genetic basis of these phenomena that are passed down from generation to generation [167].

In the phenomenon of phyllotaxis, more complex concepts of symmetry are used, in particular the concept of a "screw (helical) axis of symmetry." It should be noted that both Fibonacci numbers and the golden ratio (the golden section), which are observed in various biological materials, can serve both as characteristics of these structures and as indicators of the presence of certain non-markovian processes [168].

The Markov process is known to have discrete time and to exhibit the property of absence of memory (i.e., no feedback from past states). This characteristic of how the system's state is determined at any given moment uniquely determines the probability distribution and the future behavior of the process. Information about past behavior is not retained; only the most recent state transition is remembered. Therefore, for a Markov process, the probabilistic behavior of the system in the future is entirely determined by its state at time t_0.

$$u_{n+1} = f(u_n) \tag{6.50}$$

In non-Markov processes, a crucial additional characteristic is negentropy—the measure of the order or complexity of a structure, taking into account additional information about events that occurred before a certain time point ($t < t_0$), i.e., memories of the past. From an "informational perspective," negentropy is the antinomy of entropy; it can be considered as a concept that is "genetically" derived from entropy. Therefore, negentropy can only be understood in relation to entropy, that is, in a parallel manner. Non-Markov processes are inherently non-local in time. In the simplest case, the probabilistic structure of such processes can be represented by the equation: $u_{n+1} = un + u_{n-1}$. The value of the parameter un in this equation depends not only on the previous state but also on events that occurred before t_0. The short-term memory, as reflected in the recursive relationship (6.50), is particularly significant in many biological phenomena; this is precisely where its fundamental importance lies. The solution to the recursive equation (6.50) is the sequence of Fibonacci numbers. The Markov approach can be seen as a zero-order approximation to the general non-

Markov theory, within which one can start from successive approximations to understand the behavior of these processes.

In Section 1.4, the possibility of solving complex problems using a mathematical model based on matrix analysis is discussed. Matrix analysis has long been proven to be useful in various fields of science. In the field of genetics, this approach enables the combination of parameters from numerous individual elements into a single matrix. The symmetry of the elements in matrices gives rise to meaningful mathematical expressions; thus, matrices can be considered as objects on which certain mathematical operations can be performed. The matrices reflect the relationships between the molecular structures of genetic code.

The simplest 4 × 4 matrix is the amino acid matrix presented in Table 6.1. This table highlights the patterns observed in the symmetry of the two sets of amino acid codons. The functioning of living systems can be understood as a result of the precise adherence to these symmetry principles. Table 5.1 lists all 64 possible amino acid codon triplets; we will utilize this ordering in our matrix analysis. All these codon triplets will be represented in an 8 × 8 matrix format. In such a matrix (Table 6.19), each codon corresponds to a specific cell.

The entire sequence is strictly ordered and read from left to right and from top to bottom. Given the degeneracy of the genetic code, the same amino acid will be encoded by several different codons in various cells. Table 6.19 lists the numbers of each cell. From the total of 64 codons, numerous combinations can be obtained by rearranging them in different ways, while keeping the number of codons unchanged and only changing their order. Such combinations are known in mathematics as permutations. It is worth noting that the number of possible arrangements for the 64 codons in a matrix of this type is enormous: $64! \approx 10^{89}$. Similarly, the random distribution of the 20 amino acids and their corresponding codons also leads to a vast number of possibilities, and such distributions almost never result in any symmetry in the arrangement of the codons. The work [167] visually demonstrates an example of such a matrix filling: an (8 × 8) matrix with 64 sequentially numbered cells, 32 of which are filled with "black" and 32 with "white" numbers, selected randomly. It turns out that the resulting black-and-white pattern is highly likely to be asymmetric. Thus, one of the simplest ways to generate alternative genetic codes is to randomly distribute the 20 amino acids across the 20-codon blocks. This procedure yields 2.5×10^{18} possible configurations—roughly the number of seconds that have passed since the formation of the Earth. However, as the author [167] points out, nature has chosen a form of genetic code degeneracy that corresponds to a symmetric, and therefore regular, arrangement of the codons in the matrix. A similar symmetry can be observed in Table 6.19: in the dark cells, codons encoding a single amino acid are placed, while in the light cells, codons encoding two amino acids are located.

Table 6.19

	1	2	3	4	5	6	7	8
1	AAA1 000000	AAG2 000001	AAT3 000010	AAC4 000011	AGA5 000100	AGG6 000101	AGT7 000110	AGC8 000111
2	ATA9 001000	ATG10 001001	ATT11 001010	ATC12 001011	ACA13 001100	ACG14 001101	ACT15 001110	ACC16 001111

	1	2	3	4	5	6	7	8
3	GAA17 010000	GAG18 010001	GAT19 010010	GAC20 010011	GGA21 010100	GGG22 010101	GGT23 010110	GGC24 010111
4	GTA25 011000	GTG26 011001	GTT27 011010	GTC28 011011	GCA29 011100	GCG30 011101	GCT31 011110	GCC32 011111
5	TAA33 100000	TAG34 100001	TAT35 100010	TAC36 100011	TGA37 100100	TGG38 100101	TGT39 100110	TGC40 100111
6	TTA41 101000	TTG42 101001	TTT43 101010	TTC44 101011	TCA45 101100	TCG46 101101	TCT47 101110	TCC48 101111
7	CAA49 110000	CAG50 110001	CAT51 110010	CAC52 110011	CGA53 110100	CGG54 110101	CGT55 110110	CGC56 110111
8	CTA57 111000	CTG58 111001	CTT59 111010	CTC60 111011	CCA61 111100	CCG62 111101	CCT63 111110	CCC64 111111

For a more in-depth analysis of the properties of the genetic code, it is useful to perform various rearrangements within the matrix—for example, different permutations of the letters. These rearrangements must be identical across all 64 triplets. Since there are three letters in each cell, there are six possible ways to permute these letters. If the original order of the letters is 1 2 3, we can shift positions cyclically so that the second letter becomes first, the third becomes second, and the first becomes third—giving the new order (2 3 1).

We will make such substitutions in Table 6.19 to obtain its modified version, Table 6.20.

Table 6.20

	1	2	3	4	5	6	7	8
1	AAA1 000000	AGA5 000100	ATA9 001000	ACA13 001100	GAA17 010000	GGA21 010100	GTA25 011000	GCA29 011100
2	TAA33 100000	TGA37 100100	TTA41 101000	TCA45 101100	CAA49 110000	CGA53 110100	CTA57 111000	CCA61 111100
3	AAG2 000001	AGG6 000101	ATG10 001001	ACG14 001101	GAG18 010001	GGG22 010101	GTG26 011001	GCG30 011101
4	TAG34 100001	TGG38 100101	TTG42 101001	TCG46 101101	CAG50 110001	CGG54 110101	CTG58 111001	CCG62 111101
5	AAT3 000010	AGT7 000110	ATT11 001010	ACT15 001110	GAT19 010010	GGT23 010110	GTT27 011010	GCT31 011110

	1	2	3	4	5	6	7	8
6	TAT35 100010	TGT39 100110	TTT43 101010	TCT47 101110	CAT51 110010	CGT55 110110	CTT59 111010	CCT63 111110
7	AAC4 000011	AGC8 000111	ATC12 001011	ACC16 001111	GAC20 010011	GGC24 010111	GTC28 011011	GCC32 011111
8	TAC36 100011	TGC40 100111	TTC44 101011	TCC48 101111	CAC52 110011	CGC56 110111	CTC60 111011	CCC64 111111

It turns out that the code in question demonstrates a high degree of symmetry in the new matrix, both with regard to the arrangement of amino acids and the black-and-white mosaic pattern. The structure of this mosaic pattern features cyclic shifts.

The symmetric properties of triplet matrices also appear when alphabet permutations are applied across the full set of 64 triplets. Such permutations involve the simultaneous mutual exchange of the four letters of the genetic alphabet across all triplets, for example the mutual substitution C↔G, A↔U [167]. The symmetric properties of these new matrices are similar in nature to the permutations discussed above.

S.V. Petukhov presents the simplest square matrix of second order, P, which is used in a system consisting of four "letters" from the genetic alphabet:

$$P = \begin{bmatrix} & 1 & 0 \\ 1 & C & A \\ 0 & U & G \end{bmatrix} \tag{6.51}$$

Petukhov suggests that the family of all genetic polypeptides with identical lengths should be considered as a corresponding set of matrices, denoted by $p^{(n)}$, which represent tensor (Kronecker) powers of the original matrix. As examples, the matrices $p^{(2)}$ and $p^{(3)}$ have been constructed. The matrix $p^{(2)}$ is depicted as follows:

$$p(2) = \begin{bmatrix} & 11 & 10 & 01 & 00 \\ 11 & CC & CA & AC & AA \\ 10 & CU & CG & AU & AG \\ 01 & UC & UA & GC & GA \\ 00 & UU & UG & GU & GG \end{bmatrix} \tag{6.52}$$

The matrices of the form $p^{(n)}$ are called genomic matrices. When each symbol in these symbolic matrices is replaced by some quantitative parameter corresponding to its respective nitrogenous base, the resulting numerical genomic matrices are obtained. To derive such quantitative parameters, S.V. Petukov proposes using the numerical values of the hydrogen bonds present in the nitrogenous bases of the genetic code, which have long been recognized to carry significant informational content. These hydrogen bonds can be either two- or three-carbon bonds (in which case C = G = 3 and A = T/Y = 2). The idea behind Petukov's approach is to replace each polypeptide in all $p^{(n)}$ matrices with the product of the number of hydrogen bonds in its constituent nitrogenous bases. For example, the sequence "CC" in the genomic matrix $p^{(2)}$ is replaced by 3 × 3

= 9, and the sequence "CCC" in the genomic matrix $p^{(3)}$ by $3 \times 3 \times 3 = 27$. The numerical genomic matrix $p^{(2)}$ would thus appear as follows:

$$p(2) = \begin{bmatrix} & 11 & 10 & 01 & 00 \\ 11 & 9 & 6 & 6 & 4 \\ 10 & 6 & 9 & 4 & 6 \\ 01 & 6 & 4 & 9 & 6 \\ 00 & 4 & 6 & 6 & 9 \end{bmatrix} \quad (6.53)$$

Numerical matrices possess a variety of different properties. Firstly, they are symmetrical with respect to both diagonals; therefore, they are called "bisymmetric". Secondly, the sum of the numbers in each row and each column of the matrix is equal: in the matrix $p^{(2)}$, the sum is 25; in the matrix $p^{(3)}$, the sum is 125; the total sum of the numbers in the matrix $p^{(2)}$ is 100; and in the matrix $p^{(3)}$, the sum is 1000. It turns out that any numerical matrix of the form $p^{(n)}$ also possesses this property. Each of these matrices is symmetric, and the sum of the numbers in each row and in each column of the matrix is equal to 5n. The total sum of all the numbers in the matrix is equal to 10^n.

In the nucleotides T and U (thymine and uracil), there is no amino group present, unlike in the other nucleotides. Therefore, due to the absence of an amino group in T and U – as a distinctive characteristic (as opposed to the other three residues) – and because T and U are mutually substituted in DNA and RNA (i.e., only thymine is replaced by uracil when converting from DNA to RNA) – it is possible to correlate these nucleotides with the Hadamard matrices. Specifically, the symbols C, A, G are represented by "+1", while T/U are represented by "–1". This correlation converts all matrices of the Kronecker family $p^{(k)}$ into Hadamard matrices. The aforementioned difference between thymine and the other nucleotides almost automatically defines the pair A–T as absolute opposites, while shifting the pair C–G into a relative opposition. Together, these four bases constitute a system of four logical states, which is best encoded using a binary code (11, 10, 01, 00). Thus, in matrix genetics, the numerous genome matrices of different species are closely related to the diversity of Hadamard matrices.

Living organisms use two genetic codes to translate inherited information: the standard code for proteins, which is encoded in the cell nucleus; and a more "advanced" version of the code used for translating the mitochondrial genome.

Scientists have proposed the hypothesis that there exist multiple genetic codes, each with varying sensitivity to errors. Organisms that were better able to cope with these errors had a greater chance of survival, and consequently, the "canonical" code (the most error-resistant code) simply emerged as the winner in the struggle for survival. This hypothesis is entirely plausible, given that we now know the existence of alternative genetic codes. The first example of a deviation from the standard genetic code was discovered in 1979 during the study of human mitochondrial genes; since then, several more such variants have been found. These alternative codes are referred to as "non-standard genetic codes."

To date, 18 protein codes for mitochondria and various organisms have been identified (http://www.ncbi.nlm.nih.gov/Tahonomy/Utils/wprintgc.cgi); these are presented in Table 6.21 (according to information from the Internet).

The code is not universal. In genetic codes, there are two subsets of amino acids: one consists of 12 low-degeneracy amino acids (with degeneracy ranging from 1 to 3), and the other consists of 8 high-degeneracy amino acids (with degeneracy ranging from 4 to 8). This rule is apparently followed 100% in autotrophic organisms, which produce organic matter by obtaining carbon for their bodies through the process of photosynthesis.

For heterotrophic organisms, there is a slight exception to this rule: three out of the eighteen genetic codes have low-degeneracy and high-degeneracy amino acids in a ratio different from 12:8, namely 11:9 (code No. 6) and 13:7 (codes No. 3, 18). These non-standard ratios represent the minimum possible deviations from the canonical ratio of 12:8 (since the sum of the numbers in these codes is 20). Additionally, the numbers 8 and 12 are stereometrically related, as they are fundamental parameters defining the structure of a cube. A cube has 8 vertices and 12 edges. It should also be noted that if a single codon encodes different amino acids in various genetic codes, then those amino acids belong to the same canonical set of amino acids of high or low degeneracy [167].

The analysis revealed that codes #3, 5, 7, 12, and 15 each contain an amino acid with a maximum degeneracy number equal to 8. For codon #3, this amino acid is threonine; for the remaining codons (#5, 7, 12, and 15), it is serine.

Rather than using a single set of 20 amino acids, nature combines two different subsets of amino acids—one containing 8 amino acids and the other containing 12 amino acids. The smallest common multiple of these two subsets is 24. The primary numbers associated with the degeneracy of amino acids in all 18 genetic codes (codes #1, 2, 3, 4, 6, 8) are factors of the number 24. The number 24 thus serves as a constant that ensures the consistency of amino acid degeneracy across these codes. In biology, it is known as a number related to the division of a day into equal parts and to the way protein systems in organisms function depending on the time of day.

Table 6.21

Variants of genetic codes			
1	**2**	**3**	**4**
1	The Standard Code	12	8
2	The Vertebrate Mitochondrial Code	12	8
3	The Yeast Mitochondrial Code	13	7
4	The Mold, Protozoan, and Coelenterate Mitochondrial Code and the Mycoplasma/Spiroplasma Code	12	8
5	The Invertebrate Mitochondrial Code	12	8
6	The Ciliate, Dasycladacean and Hexamita Nuclear Code	11	9
7	The Echinoderm and Flatworm Mitochondrial Code	12	8
8	The Euplotid Nuclear Code	12	8
9	The Bacterial, Archaeal and Plant Plastid Code	12	8
10	The Alternative Yeast Nuclear Code	12	8
11	The Ascidian Mitochondrial Code	12	8

Variants of genetic codes			
1	**2**	**3**	**4**
12	The Alternative Flatworm Mitochondrial Code	12	8
13	Blepharisma Nuclear Code	12	8
14	Chlorophycean Mitochondrial Code	12	8
15	Trematode Mitochondrial Code	12	8
16	Scenedesmus obliquus Mitochondrial Code	12	8
17	Thraustochytriun Mitochondrial Code	12	8
18	Pterobranchia Mitochondrial Code	13	7

Note: 1 — code numbers; 2 — code names; 3 — sums of amino acids with degeneracy number from 1 to 3; 4 — sums of amino acids with degeneracy number from 4 to 7 (for codes No. 3 and 18), up to 9 (for code No. 6), and up to 8 (for all others).

Thus, the original question in the field of genetics concerning the origin of the 20 amino acids boils down to the issue of whether alternative sets of 8 or 12 amino acids could have existed. It has been observed that in all genetic codes except codes 9 and 10, the set of corresponding ter (stop) codons undergoes changes; the most frequent change (reinterpretation) occurs with the UGA codon (codons 2, 3, 4, 5, 7, 8, 11, 12, 15, 18). At the same time, the triplet UGA can encode the 21st amino acid, selenocysteine, which is found in some natural proteins. This requires the presence of an additional structural element in the mRNA. Selenocysteine is therefore a non-standard "21st" amino acid, differing from regular cysteine in that it contains a selenium atom instead of a sulfur atom. Selenocysteine is an essential component of several enzymes in animals (including humans), protozoa, bacteria, and archaea. Proteins containing selenocysteine are called selenoproteins. The human genome contains more than 20 genes for selenoproteins; thus, selenium is an essential nutrient, and its deficiency in the diet leads to various diseases. The UGA codon is used to encode selenocysteine, although it is normally a stop codon that signals the end of protein synthesis. However, if a specific nucleotide sequence called SECIS follows the coding region of a gene, the protein synthesis machinery interprets the UGA codon as encoding selenocysteine. The matrix method allows for the alignment of nucleotide sequences; this approach is known as "point alignment." As an example, let's consider two nucleotide sequences:

```
T A C G C A G C A A          (1)
T A G G C G C T A            (2)
```

The matrix alignment method consists of comparing the nucleotides in these two sequences.

When the nucleotides in sequences (1) and (2) are identical, dots are placed opposite them. The dots will be located on the diagonal if the nucleotides match;

otherwise, they will be shifted upward or downward if the nucleotides do not match (see Figure 6.5).

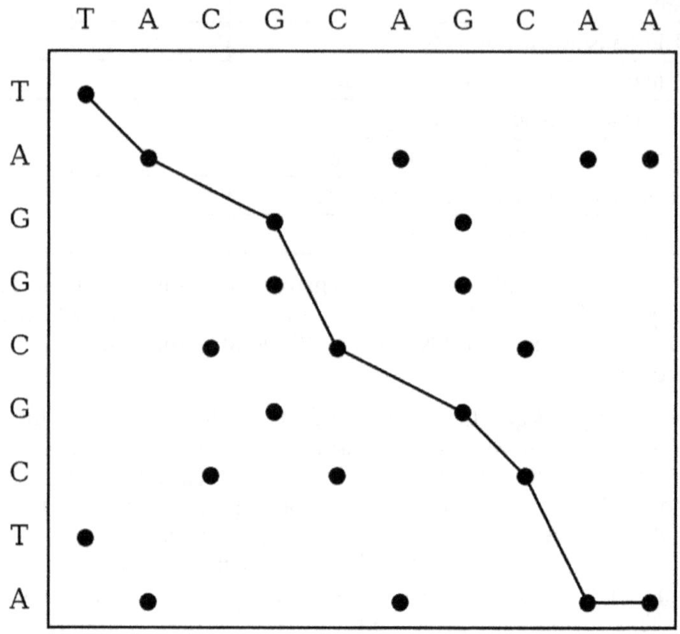

Figure 6.5

CHAPTER 7. THE LOGIC OF RECOGNIZING GENETIC INFORMATION

7.1. THE ROLE OF FEATURES IN THE RECOGNITION PROCESS AND THEIR SELECTION

To recognize a certain structure, we will need features. Among the possible features that can be used are elementary molecular features: nucleotide positions. These nucleotide positions can have four different states, depending on the specific nucleotide that constitutes them. The four nitrogenous bases – A (Adenine), G (Guanine), T (Thymine), and C (Cytosine) – represent specific polyatomic structures with unique biochemical properties. For example, two of the bases, C and T (U), are equivalent to each other based on their pyrimidine structure (they both have a single ring in their molecular structure); the other two bases, A and G, are equivalent with respect to the absence of a ring in their structure (they belong to the purine class). Based on the presence or absence of strong or weak hydrogen bonds in the complementary pairs of nitrogenous bases, another type of equivalence can be established: G = C (form a complementary pair with three hydrogen bonds), A = T (form a complementary pair with two hydrogen bonds).

When dividing the letters into equivalent pairs based on a specific feature, each letter can be associated with a symbol representing a binary opposition: for example, "1" if the letter is present in the sequence and "0" if it is absent. Nucleotide sequences, amino acids, entire genes, or proteins can also be considered as molecular features.

Mathematical-statistical methods (such as maximum likelihood and the Bayesian approach) have a number of undeniable advantages over the classical approach: they allow for the use of a much larger number of features and enable statistical evaluation of the results of the calculations.

When there is insufficient information about the chemical or biological structure, a hypothetical model is constructed. A hypothesis is a mental assumption or a logical extension of a series of judgments that provides an explanation for the nature of a phenomenon; however, the accuracy of this explanation remains approximate. Hypotheses are an integral part of every method of biological research (especially experimental research). Therefore, they constitute a necessary component of biological knowledge [169].

Hypothetical models stimulate the intellectual process of understanding the surrounding world. In the mid-17th century, statistical models began to emerge in science. To study these models quantitatively, it was necessary to develop new mathematical methods based on the theory of probability and mathematical statistics.

If the objects are described in terms of binary features and are then represented as Boolean vectors, the simplest way to build a model is to look for patterns (i.e., to identify relationships or structures among the features).

The process of identifying a pattern is an inductive one. If a hypothesis is confirmed, then that is the pattern we are looking for; if not, we need to formulate a new hypothesis and repeat the verification process. Making assumptions about the existence of patterns (hypotheses) based on a selected set of facts is essentially what induction entails. The number of possible structures of an object can be quite large, so we need to decide how to represent that object's structure. A graph is a very convenient tool for describing such structures.

We will construct two vectors from these bases:

in the form of a row vector:

$$X=[A\ G\ T\ C];$$

and a column vector:

$$U = \begin{pmatrix} A \\ G \\ T \\ C \end{pmatrix}$$

The product of these vectors will be a matrix of duplets.

$$A = \begin{bmatrix} AA & AG & AT & AC \\ GA & GG & GT & GC \\ TA & TG & TT & TC \\ CA & CG & CT & CC \end{bmatrix}$$

If each element can be represented by a number, for example: A = 0; G = 1; T = 2; C = 3, then the sum of each pair of elements can be expressed as: $d_{ij} = d_i + d_j$ And then the matrix A can be represented in digital form:

$$A* = \begin{bmatrix} 0 & 1 & 2 & 3 \\ 1 & 2 & 3 & 4 \\ 2 & 3 & 4 & 5 \\ 3 & 4 & 5 & 6 \end{bmatrix} \quad (7.1)$$

This matrix exhibits clear symmetry; its determinant is equal to zero. Therefore, this matrix is a special case. The trace of the matrix is equal to 12 and coincides with the sum of the elements on the anti-diagonal.

Obviously, multiplying the matrix A by one of the vectors will result in a three-dimensional matrix consisting of all 64 triples; in fact, this will be a three-dimensional cube. The edges of this cube will be made up of 4 triples each, and the faces will be made up of 16 triples each. The structure of such a triplet genetic code is depicted in Figure 7.3.

Based on the complete list of symmetries and numerical patterns in the individual elements of the cubic version of the genetic code, the following conclusions can be drawn:

- The algebraic approach enables us to view the problem of genetic code from a new perspective.
- The symmetries observed in the code are reflected in the process of selecting codons to determine various amino acids.

Nucleotides can uniquely determine an amino acid. The introduction of the concept of "degree of determination" for nucleotides allows us to represent genetic texts as a sequence of numbers from 1 to 4.

In [170], a four-valued logic is proposed, represented by basic functions: 0 – no; j – can never be; i – can be; 1 – yes.

V. I. Lobanov emphasizes the complementary nature of the values of the variables: 0 + 1 = 1, i + j = 1, 0 & 1 = 0, i & j = 0. Therefore, he argues that it is quite natural to refer to this type of logic as "complementary logic." For the basic functions of complementary logic (just as for the three-valued logic), De Morgan's laws hold true—these laws relate the negation operator to the conjunction and disjunction operations.

7.2. GRAPHICAL REPRESENTATION OF THE MODEL

The essence of the graphical model lies in representing multi-dimensional distributions using graphs.

The vertices of the graph are connected by arcs, which represent the relationships between those vertices. The graph can be modified; for example, you can arbitrarily add a new edge to the graph, remove a randomly selected edge, or change the graph in some other way. By repeating this process a sufficient number of times, you can be confident that you will eventually arrive at the desired graph.

7.3. IDENTIFYING PATTERNS

If a researcher clearly knows the features (or characteristics) based on which objects can be distinguished, it is not particularly difficult for them to classify those objects into the corresponding classes. For example, the difference between two classes might be that one class includes odd numbers and the other includes even numbers. If no such clear differences are apparent, then finding a combination of relevant variables that can be used to define functions determining the membership of objects to their classes can be accomplished through a systematic (but time-consuming) search. However, this systematic approach is not very efficient. Therefore, we have to rely on experts and specialists who have knowledge of the subject matter in question. Almost always, there is some intuition or expert judgment that can be used to determine the degree of similarity between a particular real object and the classes to which it should be assigned.

Let's consider an example that demonstrates the possibility of constructing a classification algorithm based on the analysis of samples.

Let the original set V represent a set of binary words of a given length n = 6. Let two samples of words be provided, belonging to two classes A and B, respectively. The samples belonging to class A are: 010111, 111001, 101101, 010001; the samples belonging to class B are: 111100, 100010, 010110, 101010.

An analysis of the contents of these two samples shows that all elements in sample A are odd binary numbers, while all elements in sample B are even binary numbers. This allows us to make a well-founded assumption: entire class A consists only of odd numbers, and entire class B consists only of even numbers. The algorithm for the desired transformation (if our assumption is correct) is straightforward: we check the lowest digit of a word; if it equals 1, the word belongs to class A, otherwise it belongs to class B. This information can be used to construct a Boolean function that distinguishes between the two classes. The function is very simple: $F(x_1, x_2, x_3, x_4, x_5, x_6) = x_6$. The degree of similarity between two words is also determined by their lowest digits: if the digits are the same, the words are similar; if they are different, the words are not similar.

Let's consider another example of samples belonging to two classes within a set of six-digit binary words. Class A consists of the words: 010101, 111001, 100110, 100101; Class B consists of the words: 110100, 100010, 010010,

101000. These samples are arranged in such a way that in Class A, there are always exactly two binary ones in each of the lower four digits, while in Class B, there is always exactly one binary one in each of the lower four digits. This arrangement allows us to construct logical functions that can distinguish between the two classes.

In this example, they are:

$$F_1(x_1, x_2, x_3, x_4, x_5, x_6) = x_3 \bar{x_4} x_5 \bar{x_6} \vee x_3 \bar{x_4} x_5 \bar{x_6} \vee x_3 \bar{x_4} \bar{x_5} x_6;$$

$$F_2(x_1, x_2, x_3, x_4, x_5, x_6) = x_3 \bar{x_4} x_5 x_6 \vee x_3 x_4 \bar{x_5} x_6 \vee \bar{x_3} x_4 x_5 x_6.$$

In this example, these are the four least-significant bit positions.

In this example, the variables that determine the membership of classes were the four least-significant bit positions.

The search for combinations of relevant variables that can be used to construct functions determining membership in certain classes can be performed using a systematic, targeted approach. Calculations show that direct enumeration is not feasible for real-world problems. Defining such functions is one of the problems in discrete mathematics [22].

We assume that the set V represents a random and equally likely sample from the set U. In practice, however, only a subset of U is actually determined as the result of an experiment; this subset encompasses all the objects belonging to the classes under study. This subset may be much smaller than U, i.e., $|V| << |U|$. But can we still solve the same problems by knowing only V and having information about a relatively small part of U? This is a fundamentally important question, as it is the case in most practical research scenarios. The set V is a binary matrix with n columns corresponding to the various characteristics (e.g., x_1, x_2, x_3, x_4, x_5, x_6) and m rows representing the selected objects (amino acids). An analogy for the set U could be a matrix with 64 rows and six columns (see Table 5.1). We assume that each element in the matrix V takes the value 0 or 1 with equal probability of 0.5. Since the number of elements in V is equal to the product of the number of rows and columns ($|V| = mn$), the total number of possible binary matrices is 2^{mn}, and all of them are equally likely. Let's consider a range of ranks, denoted by k. We define the event that the second, third, and fifth positions in a matrix from the set V contain only zeros. This corresponds to certain termination codons in amino acid sequences (e.g., TAA and TGA; see entries 33 and 37 in Table 5.1). We denote this event as i(m, n, k), where m is the number of elements in V, n is the number of characteristics, and k is the rank of the range. We want to calculate the probability of this event, denoted as p_i(m, n, k). Recalling some fundamental principles of probability theory (based on Bernoulli's theorem), we note that the probability of an event can be precisely calculated only for a complete sample—i.e., when we have data for all possible outcomes. In practice, we never have access to a complete sample; instead, we rely on smaller subsets, known as training and testing sequences. Therefore, we must use the probabilities of events calculated on the training sequence. Only in the limit of an infinitely long training sequence would these probabilities become true probabilities. Nevertheless, we are forced to work with empirical estimates (i.e., frequencies). Consider all possible binary matrices of size $m \times n$. The proportion of matrices whose second, third, and fifth positions contain zeros in each row is 2^{-k} of the total number of matrices (2^{mn}). The proportion of matrices that have at least one 1 in the second, third, or fifth position in each row is $1 - 2^{-k}$.

The probability that a specific range of ranks k contains only zeros in a randomly selected matrix from V is then given by dividing the number of favorable outcomes ($2^{mn}(1 - 2^{-k})^m$) by the total number of possible matrices (2^{mn}).

$$p_i(m, n, k) = (1 - 2^{-k})^m.$$

The a priori probability that a particular interval will be empty might be quite low. However, the probability that at least one of those intervals will be empty could be significantly higher. When calculating probabilities for complex events that are expressed in terms of simpler events whose probabilities are known, two fundamental theorems of probability theory are typically applied. One of these theorems concerns the sum of mutually exclusive events (events that cannot occur simultaneously); the probability of the sum of such events is equal to the sum of their individual probabilities. More formally stated: if A and B are mutually exclusive events, then...

$$p(A + B) = p(A) + p(B).$$

In general, when the relationship between events A and B is arbitrary, the formula is valid.

$$p(A + B) \leq p(A) + p(B).$$

Another theorem is applicable when determining the probability of the occurrence of independent events. The product of events A and B is defined as an event that occurs if, and only if, both events A and B occur. It is denoted by A·B. Events are considered independent if the occurrence of one of them does not affect the probability of the other occurring. The theorem states that the probability of the product of two independent events A and B is equal to the product of their individual probabilities:

$$p(A \cdot B) = p(A) \cdot p(B).$$

The event in question may be a rather complex outcome of several simple events, the probabilities of which are known. These simple events are interrelated in a sophisticated manner. In that case, calculating its probability boils down to first transforming the formula that represents the event in a way that expresses it as some kind of superposition of products of independent events and sums of non-overlapping events. On this path, one has to deal with complex combinatorial problems whose solutions cannot always be fully implemented in a practical manner. However, we do not need to calculate the value of the probability pi(m, n, k) for all possible values in the parameter space of m, n, k. It's sufficient to estimate this probability when its values are small. It turns out that the value of pi(m, n, k) can be well estimated using the mathematical expectation M(m, n, k) of the number of objects of rank k that do not intersect with a random set V. As is well known, the mathematical expectation of a random variable is the value that results when all possible outcomes are averaged, taking into account their probabilities through appropriate weighting [57]. In this case, all possible implementations (i.e., the various Boolean matrices of size m x n) are equally likely, which simplifies the calculation of the mathematical expectation. Consider, on the one hand, the set of all 2^{mn} matrices of a given size, which are all binary matrices; and on the other hand, the set of all objects of rank k in the space U. The number of such intervals is equal to Ckn 2k. The first factor indicates the number of different combinations when selecting k traits from a total of n traits, while the second factor represents the number of fixed combinations of values for the selected traits. Meanwhile…

$$C_n{}^k = \frac{n \cdot (n-1) \cdot \ldots \cdot (n-k+1)}{1 \cdot 2 \cdot \ldots \cdot k}.$$

Let's now consider the product of these sets, which is formed by all possible pairs of elements: the first element of each pair is chosen from the set of matrices, and the second element is chosen from the set of objects. Let's use the number of different matrices that have been calculated earlier; these matrices do not intersect with a specific object of rank k (which has been arbitrarily chosen). The value of k is equal to $2^{mn}(1-2^{-k})^m$. By multiplying this value by the number of different objects of rank k, we obtain $2^{mn}C^k_n \cdot 2^k(1-2^{-k})^m$. By then dividing this result by the number of different matrices, we can find the value of the mathematical expectation that interests us:

$$M(m, n, k) = C^k_n 2^k (1-2^{-k})^m.$$

It is evident from the way the value of M(m, n, k) is calculated that it equals the sum of the probabilities of the events i(m, n, k) for all possible objects of rank k. Since the sum of the probabilities of any set of events cannot be less than the probability of the combination of those same events, the following relationship holds:

$$pi(m, n, k) \leq M(m, n, k).$$

To get a clearer understanding of the value of M, let's calculate its value for our specific case using a pocket calculator, where m = 200, n = 6, and k = 3. The result will be:

$$M(m = 200, n = 6, k = 3) = C_6{}^3 \cdot 2^3(1 - 2^{-3})^{200} = \frac{6 \cdot 5 \cdot 4}{1 \cdot 2 \cdot 3} \cdot 8 \cdot (0.875)^{200} = 0.5 \cdot 10^{-10},$$

where m represents the number of amino acids in the protein chain.

For the number of intervals in the range where k = 4 (in this case, for the terminating codons TAA and TGA, in addition to positions two, three, and five, the position six corresponding to zero should also be considered), the value of the mathematical expectation will be equal to:

$$M(m = 200, n = 6, k = 4) = C_6{}^4 \cdot 2^4(1 - 2^{-4})^{200} = \frac{6 \cdot 5 \cdot 4 \cdot 3}{1 \cdot 2 \cdot 3 \cdot 4} \cdot 16 \cdot (0.9375)^{200} = 0.24 \cdot 10^{-5}.$$

For the number of intervals in the rank range k = 5, the value of the mathematical expectation will be equal to:

$$M(m = 200, n = 6, k = 5) = C_6{}^5 \cdot 2^5(1 - 2^{-5})^{200} = \frac{6 \cdot 5 \cdot 4 \cdot 3 \cdot 2}{1 \cdot 2 \cdot 3 \cdot 4 \cdot 5} \cdot 32 \cdot (0.96875) = 0.01.$$

There is a clear and strong dependence of M on k. When m = 200 and n = 6, and the rank of the interval does not exceed 4, the value of the mathematical expectation is very small. Consequently, since $R_i \leq M$, the probability of finding an interval with a rank not exceeding 4 that does not intersect with the set V (randomly chosen from the space U) will also be small. Therefore, if such an interval is indeed found, it would be justified to propose the hypothesis that there exists a corresponding regularity reflected in this empty interval. However, if k > 4, the discovery of an empty interval with rank k alone cannot serve as evidence for the hypothesis of a corresponding implied regularity that connects certain characteristics.

7.4. EXAMPLE OF RECOGNIZING THE SPATIAL STRUCTURE OF A PROTEIN USING LOGICAL METHODS

It was noted that the properties of homogeneous Markov chains are entirely determined by the initial probability vector and the transition probability matrix. In some cases, an oriented graph is used in place of the transition matrix; the vertices of this graph represent the states of the chain, and the edges indicating transitions from one state to another show the probability of transitioning from state E_i to state E_j (with probability P_{ij}). Therefore, graphs are proposed for the analysis of protein chains. The sequence of amino acids can be viewed as a system that is capable of changing its state. It is important to emphasize the significance of studying protein sequences in terms of the stability of states within a particular spatial structure. Interaction with the system as an object is only possible if the effects exerted on it are recognized; such influences can take the form of either a single input signal or a combination of multiple signals. The changes in the system's output signals in response to these inputs determine its overall behavior. If transition processes are not taken into account, and given constant external conditions, the output signals can be considered stable or static.

The operation of most analytical methods for studying systems and objects is based on the use of state graphs. A state graph consists of a multitude of nodes and the connections between them. The development of any system or object begins with the identification of the external influences that it will be subjected to, as well as the object's responses to those influences. Based on the functionality that the device being developed is required to perform, an appropriate technical solution is chosen in order to create a more robust model. Using this model, the effectiveness of the selected solution is tested. The recognizability of the input influences by the object is also evaluated. If an error is detected, a more thorough analysis of its behavior is conducted, taking into account all the details and nuances related to the internal implementation of the object's model. Once the error is corrected, the testing of the object is resumed from where it was interrupted. Statistical characteristics provide a good description of the behavior of digital devices when interacting with them. To achieve the desired functionality, a series of pre-designed functional logic components are incorporated into the object. As a result, each of the implemented functions represents a specific functional block; the components within this block transform the input signals according to the applied logic into the desired output signal or some internal state. The following logical operations are used: disjunction (OR), conjunction (AND), and negation (NOT). The Boolean function for each of these logical operations will be defined using truth tables (Table 7.1).

Table 7.1

Disjunction	Conjunction	NOT operation
0+0 «-» 0	0+0 «-» 0	x | f
0+1 «-» 1	0+1 «-» 0	—
1+0 «-» 1	1+0 «-» 0	0 | 1
1+1 «-» 1	1+1 «-» 1	1 | 0

Any logic can be used to encode the graph, but the physical operations used to reconstruct the encoded structure must be consistent with that logic.

In the statistical analysis of the genetic code, it was observed that the first two bases in a triplet have the most significant influence on the properties of the corresponding amino acid, while the third base has a relatively weak effect. Based on this finding, Table 6.1 was constructed. The table includes 16 pairs of nucleotide triplets, where the first two bases are each associated with a specific amino acid:

- AA(0000) → Lysine (Lys), Asparagine (Asn);
- AG(0001) → Arginine (Arg), Serine (Ser);
- AT(0010) → Isoleucine (Ile), Methionine (Met);
- AC(0011) → Threonine (Thr);
- GA(0100) → Glutamate (Glu), Aspartate (Asp);
- GG(0101) → Glycine (Gly);
- GT(0110) → Valine (Val);
- GC(0111) → Alanine (Ala);
- TA(1000) → Stop, Tyrosine (Tyr);
- TG(1001) → Stop, Tryptophan (Trp), Cysteine (Cys);
- TT(1010) → Leucine (Leu), Phenylalanine (Phe);
- TC(1011) → Serine (Ser);
- CA(1100) → Glutamine (Gln), Histidine (His);
- CG(1101) → Arginine (Arg);
- CT(1110) → Leucine (Leu);
- CC(1111) → Proline (Pro).

This simplification facilitated the development of a functional model and a diagram depicting the relationships between these nucleotide triplets and amino acids. The graph representing these relationships can be visualized as a sphere consisting of two hemispheres, as illustrated in Figure 7.1(a) and (b).

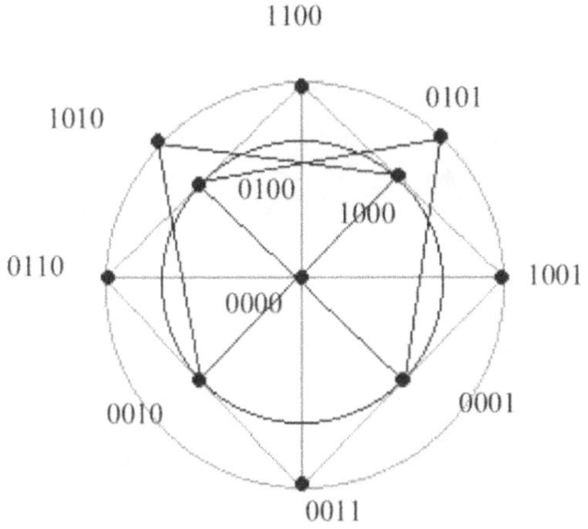

Figure 7.1(a): Projections of nodes and edges from the northern hemisphere

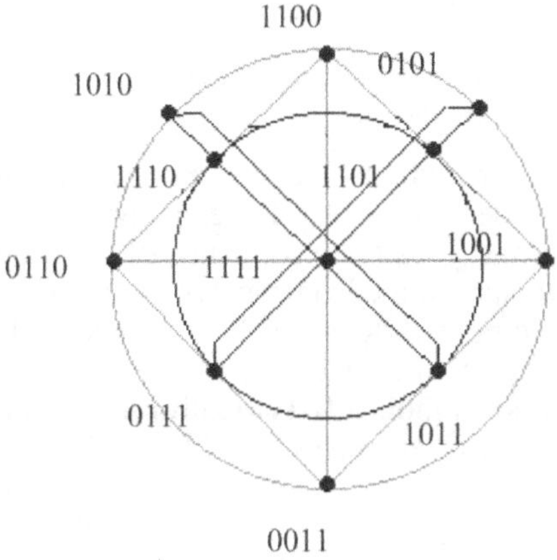

Figure 7.1b: Projections of the nodes and edges of the southern hemisphere

The position of each point in space can be determined by the positions of its nearest neighbors. The neighboring variables differ from each other by one bit. For example, the vertex 0000 is connected to the vertices 1000, 0100, 0001, and 0010.

The relationship graph can also be represented as a four-dimensional Boolean hypercube called B_4, which will have the following appearance (see Figure 7.2):

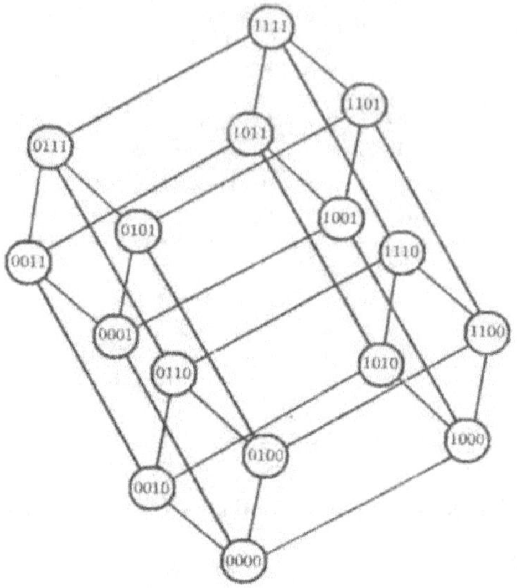

Figure 7.2. A graph depicting the relationship between four variables, represented as a Boolean hypercube with the label B_4

Based on the given transitions, we can construct a structure for the doublet genetic code that resembles the Boolean hypercube B_4. In this case, mathematicians would say that the structure of the doublet genetic code is isomorphic to the structure of the Boolean hypercube B_4.

The essence of combinatorics, which is represented by a cube (a three-dimensional representation of a square), symbolizes the final stage of a cyclical process; therefore, it is symbolically (virtually) incorporated into a circle. The meaning of the cubic form represents stability, static perfection, and truth that remains unchanged throughout time and space—i.e., it is true at all times, in all places, and from every perspective. Moreover, the sides of the cube can embody the sacred cycles of existence represented by [39].

The mathematical definition of graph isomorphism states: "Two graphs are called isomorphic if there exists a one-to-one correspondence between their vertices and edges, such that the corresponding edges connect the corresponding vertices." Following this strict definition, the work [104] indicates that the isomorphism observed is not entirely complete. The difference between these two structures lies in the genetic code: additional diagonal lines (rarely used, dashed) are present, indicating the transitions from C <--> A and G <--> U in the genetic sequence (of the type α). Such structures with additional elements superimposed on them are called simplices.

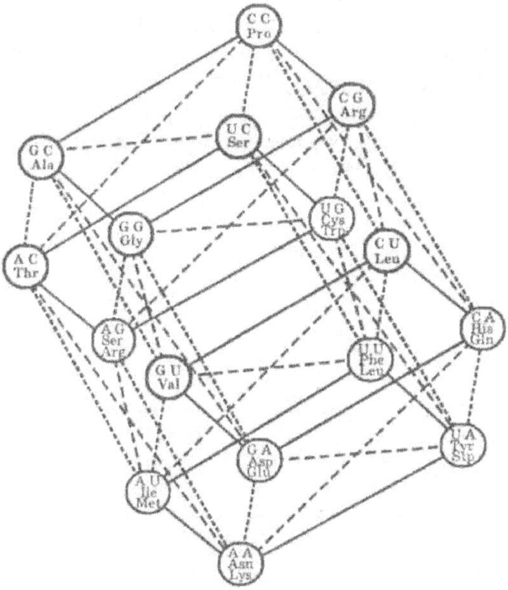

Figure 7.3. The structure of the doublet genetic code, which is isomorphic to the Boolean hypercube B_4

In analyzing the arrangement of amino acids within the double-stranded genetic code, the author [104] observes that the sequences AC, UC, AG, and UG form a cycle. This cycle contains all amino acids with C–O and C–S groups (Serine, Threonine, and Cysteine). The doublets GG, CG, UG, and AG also form a cycle and encode all Arginine (Arg) residues. The cycle consisting of GU, CU, UU, and AU contains only non-polar amino acids (Valine, Leucine, Isoleucine, Phenylalanine, Methionine), and it includes all Leucine residues as well. Finally, the cycle with the sequences AA, GA, CA, and UA contains only polar amino acids (Asparagine, Lysine, Aspartic Acid, Glutamic Acid, Histidine, Glutamine, Tyrosine). Knowing the structure of these amino acids, it is very easy to establish a pattern of change in the properties of their side chains when the nitrogenous bases at positions one and two are replaced.

Let's take a look at the structure of the triplet code.

Table 5.1, which is presented in Chapter 5, allows for quick identification of the amino acids corresponding to a particular triplet sequence.

A group of Mexican researchers led by Miguel Jiménez-Montano proposed a method for spatially representing triplets and their relationships. The publication describing this method was published in 1996 in the journal "BioSystems." They characterized each base using two independent properties.

1. By chemical class: R-purines (A, G), Y-pyrimidines (T, C);
2. The strength of the hydrogen bond: The letter "W" represents the weak hydrogen bond in the Watson-Crick base pair A=T, while the letter "S" represents the strong hydrogen bond in the G≡C base pair.

After examining these properties, the authors concluded that they could be represented by a binary number that takes either the value 0 or 1: the first digit would indicate the chemical type (R or Y), and the second digit would indicate the strength of the hydrogen bond (W or S). As a result, they assigned the following binary codes to the nitrogenous bases: A = 00; T = 10; G = 01; C = 11. For the triplet code, there are six such variables. The authors proposed a spatial structure for this triple-code representation that is isomorphic to the Boolean hypercube B_6 (Figure 7.4).

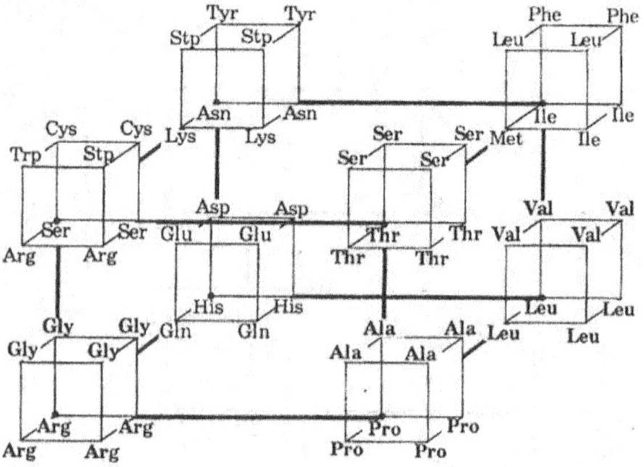

Figure 7.4. The structure of the triple genetic code, which is isomorphic to the Boolean hypercube B_6 (according to Jimenez-Montano et al.)

In the hypercube, lines connect variables that differ from each other by exactly one bit of information. To study the connections and dependencies within the hypercube B_6 in more detail, special approaches will likely be required.

Let's now look at an example of recognition that involves six features (or "signals"/"traits").

Suppose that amino acids are also used as targets for recognition. Further assume that during the training phase, 60 amino acids were observed (which corresponds to 180 nucleotides), resulting in the creation of Table 7.2. Some rows in this table may contain identical values.

Table 7.2

	X_1	X_2	X_3	X_4	X_5	X_6
Amino Acid 1	1	1	1	1	1	0
Amino Acid 2	1	1	0	1	0	1
…………………						
Amino Acid 60	1	0	0	0	1	0

The table is not reproduced in its entirety to avoid cluttering the text. The sequence of these amino acids was presented in Table 6.8. Therefore, the set of all amino acids is defined by a matrix consisting of 60 rows and 6 columns. This matrix will be used as a model for the study of amino acids. Each nucleotide in the triplet is corresponding to a pair of binary values. Specifically, the first nucleotide corresponds to the binary variables x_1 and x_2, the second nucleotide to x_3 and x_4, and the third nucleotide to x_5 and x_6.

During the analysis of the entire matrix, it was discovered that a number of amino acids are absent when considering all possible combinations of nucleotides. Specifically, the following triplets are not present in the matrix: ACA, ACG, ACT, ACC, ATA, ATC, GGA, GGC, TTC, TCA, TCG, CAA, GTA, GCG, GTG, GTC, CAT, TAA, TGA, TGT, TAC, TGC, CAC, CGC, CTC, CCC, CGA, CGT, CCA, and CCG. These triplets will be considered as "forbidden regions."

This model can also be specified in an algebraic form using a Disjunctive Normal Form (DNF) representation for the constraints. For the given example, the DNF representation looks as follows:

φ = ACN ∨ ATA ∨ ATC ∨ GTX ∨ GTC ∨ GCG ∨ GGA ∨ GGC ∨ CAM ∨ CGM ∨ CYC ∨ CCX ∨ TCX ∨ TTC ∨ TGC ∨ TAA ∨ TAC ∨ TGA ∨ TGT,

where N = A, G, T, C; M = A, T, C; Y = T, C; X = A, G.

All the amino acids listed in Table 5.1, together with the "forbidden zone," will represent the entire Boolean space of M. This Boolean space can be represented by Table 7.3, which consists of 64 cells (8 rows and 8 columns).

Table 7.3

	1	2	3	4	5	6	7	8
1	AAA 000000	AGA 000100	AGT 000110	AAT 000010	AAC 000011	AGC 000111	AGG 000101	AAG 000001
2	ATA 001000	ACA 001100	ACT 001110	ATT 001010	ATC 001011	ACC 001111	ACG 001101	ATG 001001
3	GTA 011000	GCA 011100	GCT 011110	GTT 011010	GTC 011011	GCC 011111	GCG 011101	GTG 011001
4	GAA 010000	GGA 010100	GGT 010110	GAT 010010	GAC 010011	GGC 010111	GGG 010101	GAG 010001
5	CAA 110000	CGA 110100	CGT 110110	CAT 110010	CAC 110011	CGC 110111	CGG 110101	CAG 110001
6	CTA 111000	CCA 111100	CCT 111110	CTT 111010	CTC 111011	CCC 111111	CCG 111101	CTG 111001
7	TTA 101000	TCA 101100	TCT 101110	TTT 101010	TTC 101011	TCC 101111	TCG 101101	TTG 101001
8	TAA 100000	TGA 100100	TGT 100110	TAT 100010	TAC 100011	TGC 100111	TGG 100101	TAG 100001

In this table, each amino acid (each specific combination of nucleotides) corresponds to a separate cell. The arrangement of the amino acids, and consequently the values assigned to the Boolean variables (6 variables for each amino acid), can be determined using various combinations. To ensure that the amino acids in adjacent cells differ by no more than one variable, it is advisable to use the Gray code. The Gray code assigns a unique binary representation to each natural number starting from zero; the number of bits in this binary representation depends on the number of amino acids being encoded. In this way, the most similar amino acids are encoded by the closest binary codes. If an error occurs during the transmission of the code sequence, it will result in minimal distortion of the information. Therefore, the Gray code minimizes the likelihood of errors. Table 7.3 lists all the possible triplets of nucleotides along with their corresponding six-digit values represented as Gray codes. Based on Table 7.3 and the associated amino acid values, the desired "restriction function" can be implemented.

Suppose it has been established that the amino acid under investigation, i.e., its triplet code, corresponds to the following features: the first nucleotide is T(10), the second nucleotide is C(11), and the third nucleotide is unknown. What can we say about the other features, namely the values of x_5 and x_6, which correspond to the third nucleotide in the triplet code? Let's refer to Table 7.3. Clearly, for the first and second values in the triplet code (TC(1011)), the identified amino acid can only be found in the seventh row of the table; therefore, we can limit our consideration to that row alone. By highlighting that row, we obtain a portion of Table 7.3:

1	2	3	4	5	6	7	8
TTA Leu 101000	—	TCT Ser 101110	TTT Phe 101010	—	TCC Ser 101111	—	TTG Leu 101001

The values in the corresponding columns indicate prohibited areas; therefore, the recognized amino acid cannot correspond to the prohibited (marked) cells. Among all the permissible cells, there are two cells that have the nucleotides TC(1011) in both the first and second positions. These cells are located at the intersection of row 7 and columns 3 and 6, and they correspond to the triplets TCT and TCC, which represent the amino acid serine. This is how the amino acid is identified based on the observation results.

We will analyze the basic conjunctions of various ranks.

There is a relationship between conjunctions. Such relationships are called implicative because they correspond to the logical notion of "if…, then…". The strength of an implicative relationship increases as its rank decreases. Clearly, the rank cannot be zero, as that would imply that all objects were prohibited. We also exclude relationships of rank one as being trivial. Therefore, among the non-trivial relationships, those of rank two are the strongest. Each of these relationships connects two features (in our case, x_i and x_j). We will begin by analyzing the elementary conjunctions of rank two. In our example, a relationship of rank two is only found between the features of the two columns; for each pair of columns (out of the six available), all 15 possible combinations are considered, and for each combination, the four possible elementary conjunctions (00, 01, 10, 11) are examined.

There are a total of 15 such combinations:

001111 (ACC), 011011 (GTC), 011101 (GCG), 010111 (GGC), 110110 (CGT), 110011 (CAC), 110111 (CGC), 111100 (CCA), 111011 (CTC), 111111 (CCC), 101011 (TTC), 101101 (TCG), 111101 (CCG), 100000 (TAA), 100111 (TGC).

Next, we examine the combinations of three columns and identify those that do not contain any specific sequence patterns. In this way, we find another 9 relationships :

001011 (ATC), 011001 (GTG), 001101 (ACG), 001110 (ACT), 100011 (TAC), 100110 (TGT), 110010 (CAT), 110100 (CGA), 101100 (TCA).

Then, we consider the combinations of four columns:

001100 (ACA), 010100 (GGA), 011000 (GTA), 100100 (TGA), 110000 (CAA).

Finally, we examine the combinations of five columns:

001000 (ATA).

In this way, we have identified all the cells containing amino acids that are prohibited from appearing in certain sequences. There are a total of 30 such prohibited combinations. The set of all these elementary conjunctions describes the entire area of prohibition. The remaining 34 allowed cells correspond to the 60 amino acids listed in Table 6.8. This is because some amino acids appear multiple times within the sequence. As previously mentioned, it is possible to determine the homology or partial homology between different sequences; two sequences are considered homologous if they share a common evolutionary origin. Therefore, these results may relate to the structure of the same gene in different organisms or to related genes within the same organism. The logical analysis outlined above can be used to compare sequences with each other or a sequence with a group of others, including implicative relationships.

Since we used Gray's code, let's talk about it in more detail. Gray codes were proposed in 1953 in response to engineering demands for constructing angle-to-code transformations optimal by the criterion of minimum non-uniformity error. A key feature of Gray's codes is that, in a binary space (or binary number system), when transitioning from the representation of one number to the representation of its neighboring larger or smaller number, the digit changes (from 1 to 0, or vice versa) only in one single bit position.

If there is a binary representation of a number, to represent the number that is one greater than it, simply replace the first zero from the rightmost side with a 1, and replace all the ones that follow the first zero with zeros. For example:

A = 101011(2) = 43(10);

A = 101100(2) = 44(10).

In regular binary code, when transitioning from the representation of a number to the representation of its neighboring larger or smaller number, changes to the digits may occur across multiple bits. For example, when transitioning from the number 7 (binary code 0111) to the number 8 (binary code 1000), all four bits change simultaneously. This can lead to significant errors in certain methods of encoding continuous messages in binary format.

To convert a binary positional code to a Gray code (direct left-hand (Gray code)), you need to sum the corresponding bit from the same position in the original code and the adjacent bit from the next higher position, taking the result modulo 2. The simplest way to perform this conversion is to add the binary code to itself, but shifted one bit to the right. In this process, the least significant bit of the shifted number is discarded, and the sum is calculated modulo 2 (i.e., without considering any carry bits).

As an example, consider the conversion of the binary number 1000110 to its Gray code in lexicographic order (left-to-right):

```
  1 000 110
+   100 011
  1 100 101
```

When generating the direct Gray code, the leftmost bit of the highest-order (leftmost) rank of the Gray code matches the value of the leftmost bit of the binary position code [171].

In addition to Gray codes, there are also error-correction codes that enhance reliability in the presence of random disturbances. Depending on their capabilities, these codes can be classified as either error-detecting or error-correcting. Error-detecting codes allow for the identification of errors, while error-correcting codes can not only detect errors but also determine which specific code elements have been corrupted. This is sufficient for correcting errors, as error-correcting codes are designed to work in a binary format; to fix an error, one simply needs to replace the corrupted bit with its opposite value (0 to 1 or 1 to 0). The greater the number of prohibited combinations relative to the allowed ones, the stronger the code's error-correction capability—however, this also leads to increased redundancy. In a decoding device, each received code combination is compared to the set of allowed combinations; if a received combination does not match any of the allowed ones, it is considered an indication of an error. By comparing the received combination with the allowed ones, the closest matching combination can be identified and deemed correct, thereby "correcting" the error. The most widely used codes are the "multiple" codes. In multiple codes, all transmitted code combinations are represented as binary numbers that divide evenly by a predetermined check number specific to that code. If there is a remainder after dividing the received number (represented by the code combination) by this check number, it indicates that the combination has been corrupted. Multiple codes include what are known as cyclic codes; the development of cyclic codes represents the result of searching for the simplest possible method of code transformation. Cyclic codes are constructed from ordinary digital codes through specialized logical processing.

The mathematical basis for constructing cyclic codes lies in representing any binary number as a polynomial that includes a fictitious variable x, where the binary digits serve as coefficients. An example of how to construct a cyclic code is provided in the work [10].

CHAPTER 8. PROBLEMS AND APPROACHES IN GENETIC RESEARCH

After obtaining information about the gene sequence and gene mapping, biologists realized that much more data is needed in order to understand the various processes occurring within living cells. Geneticists in several countries are conducting larger-scale studies, similar to the "Human Genome Project." Most of the researchers working on genome sequencing currently aim to achieve an error rate of no more than 1 in 10,000 nucleotide base pairs; in some cases, a precision of 1 in 100,000 is considered necessary. However, individual variations in the genome average out to about 1 in 500 base pairs. Therefore, for a complete genome sequencing project, an error rate of 1 in 1,000 is considered a more appropriate estimate of acceptable accuracy. To enhance reliability and identify potentially significant individual differences, the most biologically or medically important regions of the genome should be studied more thoroughly. Nevertheless, using a less stringent standard for other parts of the genome significantly reduces the overall cost of sequencing the human genome [172]. In this regard, high-throughput data analysis can provide substantial assistance in these research efforts.

Genetic science possesses a vast amount of information in the form of a database. A database is a collection of data that is organized in a way that facilitates easy and rapid retrieval of the desired information (http://www.answers.com/topic/database).

And yet, to this day, genetics does not have a firm consensus regarding the number of genes contained within the human genome. One of the challenges in genetics is the difficulty of identifying the interactions between genes and proteins that determine their respective functions.

The genetic code has posed a number of problems:

- Does the genetic dictionary have any physical or molecular meaning? Or is the correlation between codons and amino acids completely random?
- What can be said about the evolution of the code in this context?
- What factors affect the reading of code, as well as the processes of transcription and translation?
- What is it that causes the code to be corrupted/damaged, and how exactly?
- What are the physicochemical reasons for mutations?

The problem with the genetic code has a clear informational meaning.

There are three main directions in the development of research efforts aimed at deciphering the functional properties of living systems at the atomic level.

The first step involves intensifying classical methods of function analysis and applying these methods at the whole-genome level. Classic examples of such approaches include the study of microchips to analyze the RNA content within cells, or the simultaneous analysis of multiple protein interactions. This approach is undergoing rapid development, and its significance cannot be overstated.

The second approach, which has already received initial impetus, deserves more attention. It involves creating computer models of living cells that integrate all existing knowledge and demonstrate how the various components of a cell function in coordination with each other. The first step in this direction has been taken: an international consortium was established with the goal of developing a complete computer model of E. coli [173]; this represents a logical continuation of the genome-wide research strategies being pursued in the study of living systems.

The third approach is an experimental study aimed at identifying the smallest possible cell structure. This approach will be particularly pursued by the renowned company TIGR, which has already made significant contributions to the elucidation of the structures of numerous genes. Ultimately, all these approaches should provide a comprehensive and reliable understanding of the functioning of living systems at the cellular level.

These studies will represent the main breakthroughs in biological research in the 21st century. The biology branch that conducts these studies will become a science that integrates a vast amount of knowledge and methods from mathematics, physiology, molecular biology, physics, nanotechnology, and other fields.

It should be noted that biological processes work the same way in different organisms. Many organisms, such as yeast or E. coli, have been well-studied and can be used as models in simplified experiments.

The bacterium Escherichia coli (E. coli) is one of the best-studied organisms; virtually all information regarding its genetics, biochemistry, and molecular biology has been obtained. The natural habitat of this bacterium is the human intestine.

V.A. Ratner, in his research on the bacterium E. coli [102], established the principle of the connection between different coding sequences. According to the author, the primary characteristic of living organisms is their ability to reproduce themselves, that is, to synthesize all their macromolecular components and to pass them on by heredity. These properties are ensured within cells by the presence of an MGSU (Molecular-Genetic Control System), which is composed of coding macromolecules.

Usually, groups of equivalent codons (corresponding to the same amino acid) are called series.

The two series (methionine and tryptophan) are not degenerate; each contains only one codon. Of the 18 degenerate series, 17 have the connectivity property, meaning they form tight groups. As a result, it is possible to transition from any codon in a connected series to any other synonym through a series of sequential substitutions. If such a transition can be accomplished in a single step, the series is considered "fully connected." The connectivity property is only violated in one series (Ser); even in this case, it breaks down into two connected sub-series. This connectivity may have historical origins. If we assume that the scope of codon influence expanded at the time of code formation, then one codon could have transformed into another simply because a single letter was randomly replaced in the anticodon region of the tRNA.

However, degeneracy can be of various types. Degeneracy is called systematic if the synonymous codons differ in the third position; this difference can occur either between purines (R = A or G) or pyrimidines (Y = T or C), or simply between any of the four nucleotides (N = A, G, T, or C). Thirty pairs of codons, as well as 8 out of a total of 16 possible tetrad combinations, satisfy these criteria. All these pairs are connected, and the tetrads are fully connected. The remaining variants of degeneracy are called non-systematic. They belong to large series: Leu and Arg are connected series; Ser is an unconnected series; Ile is a three-codon fully connected series.

Rumer Y.B. designated the positions of nucleotides using the notation x–y–z, where x represents the prefix, y the root, and z the suffix; the combination xy constitutes the "base" of the codon. These terms proposed by Rumer Y.B. reflect analogous concepts from linguistics: the roots of words determine their meaning. Any mutation affecting the root (y) of a codon also changes the codon series, i.e. disrupts the meaning of the codon. Word prefixes also contribute to meaning, albeit less significantly; many changes in prefixes can change the meaning of a word. Changes in the prefix (x) of a codon often alter its meaning as well, though sometimes they result in synonymous codons. Word suffixes are typically involved in word formation (i.e., in creating synonyms). Similarly, 70% of mutations in the suffix (z) of a codon are also synonymous. Together, the prefix and root form the "base" of a codon, which carries the primary or dominant semantic load. Codons are characterized by the properties of their bases and roots. A base is considered "strong" if it completely determines the meaning of the corresponding amino acid; for example, the bases CT, GT, TC, CC, AC, GC, GG all fall into this category. It's easy to see that these bases contain nucleotides in the ratio of C: G: T: A = 7: 5: 3: 1. A base is considered "weak" if the meaning of an amino acid requires the participation of a specific third nucleotide (z) in the codon; examples of weak bases include TT, AT, TA, CA, AA, GA, TG, AG. These bases contain nucleotides in the ratio C: G: T: A = 1: 3: 5: 7.

In Table 8.1, the dark cells contain strong bases, while the light (unhighlighted) cells contain weak bases.

It is known that DNA regions with a higher abundance of G≡C base pairs are more stable than those rich in A=T base pairs. Therefore, the stronger bases form on average more hydrogen bonds with the anticodons of t-RNA molecules than the weaker bases do.

Table 8.1

AA (Lys, Asn)	AG (Arg, Ser)	AT (Ile, Met)	AC (Thr)
GA (Glu, Asp)	GG (Gly)	GT (Val)	GC (Ala)
TA (Stop, Tyr)	TG (Stop, Trp, Cys)	TT (Leu, Phe)	TC (Ser)
CA (Gln, His)	CG (Arg)	CT (Leu)	CC (Pro)

Statistical analysis shows a correlation between the GC content of genes and the percentage abundance of individual amino acids. For some amino acids, this correlation is positive—for example, with alanine—but for others, it is negative—such as with lysine. As the correlation analysis of these relationships demonstrates (Figure 8.1), all amino acids can be divided into three groups:

- There is no reliable correlation between the GC composition of the genes (the corresponding columns are shaded in gray).
- They show a reliable positive correlation with the GC composition (not crossed out).

• The data shows a reliable negative correlation (highlighted in black).

For an amino acid sequence to fold into a water-soluble, functional globular protein, a certain minimum set of structural properties is required. The most important among these properties are six: the presence of hydrophobic, acidic, and basic amino acids; the presence of amino acids capable of forming α-helices, β-sheet structures, and β-turns [Ikehara K., Yoshida Y., 1998; Ikehara K., 2002, 2005, 2009; Ikehara K. et al., 1996, 2002].

The facts obtained indicate that for a significant portion of amino acids, their prevalence is strongly correlated with the GC content (the proportion of guanine and cytosine in the DNA) of the genes. However, for the amino acids that possess the six key properties required for the formation of functional globular proteins, there is no correlation with the GC content at all. By analyzing the sequences of seven prokaryotic genes, K. Ikehara demonstrates that the characteristics of the modern universal genetic code allow for a wide range of genomic GC content variations. For the amino acids that possess all six of these key properties, their prevalence is virtually unrelated to the GC content of the genes.

Amino acids possess two main characteristic properties that are crucial for the spatial structure of globular proteins: size (M = small, C = medium, B = large) and polarity/non-polarity (P/NP).

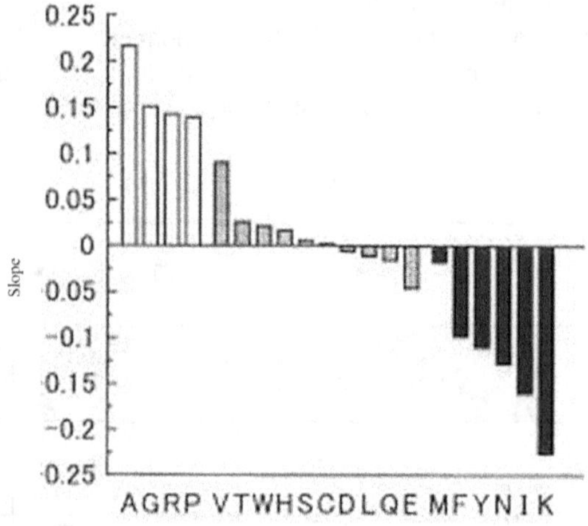

Figure 8.1. Each amino acid is represented by a one-letter code [Ikehara K., 2002]

However, it has been discovered that in the human genome, in addition to genes, there are also pseudogenes. Some of these pseudogenes contain introns, just like genes do; others lack introns entirely or only partially. Many of these pseudogenes are transcribed, but they do not result in the synthesis of functional proteins (such as [174]).

Stephen Hall in [175] notes that the significant "empty regions" in the genome, which do not encode proteins and were previously considered to be "junk", actually play an important regulatory role.

Based on the work of the English biologist J.B.S. Haldane, it is evident that "junk" DNA is essential for understanding evolutionary theory. According to Haldane, without junk DNA, evolutionism would be unable to provide a mathematical explanation for how evolution actually occurs.

Numerous studies conducted by scientists from different countries in recent years provide evidence that so-called "junk" DNA plays important roles within the genome. These roles include participating in the packaging of DNA, regulating gene activity, and synthesizing non-coding types of RNA—such as transfer RNA, ribosomal RNA, and small nuclear RNAs—which perform various functions within living cells. Researchers from the Medical Center at the University of Pennsylvania (USA) have discovered that introns, or "junk" DNA sequences, are essential for the formation of ion channels in the membranes of nerve cells, which is necessary for the transmission of nerve impulses. The theory of "junk" DNA is becoming increasingly clear, with fewer and fewer blank spots remaining.

When examining the possible pairs of amino acids in protein sequences, it is found that out of the four hundred possible combinations, 23 are missing. The missing combinations primarily consist of rare amino acids. Therefore, it is highly unlikely that the absence of these dipeptide sequences indicates some kind of "prohibition" – that is, they are not related to code overlap or other similar reasons. The missing pairs include: Arg–Asp, Asn–His, Cys–Phe, Glu–Ile, His–Arg, His–Asp, His–Gly, His–Trp, Ile–Tyr, Met–Cys, Met–His, Met–Thr, Met–Trp, Met–Tyr, Phe–Met, Phe–Trp, Thr–His, Trp–Asn, Trp–Glu, Trp–His, Trp–Met, Trp–Phe, Trp–Tyr [185].

We have already mentioned that the term "secondary structure" refers to the regular conformations of the polypeptide backbone in a protein molecule.

Based on their secondary structure, proteins can be divided into two classes. The polypeptide chains in these proteins can be fully extended in length and stabilized by hydrogen bonds that connect different segments of the chain. This results in a layered structure. The formation of such structures is facilitated by the presence of a large number of glycine and alanine residues within the chain. A second, much more important type of regular organization of the polypeptide chain arises from the formation of hydrogen bonds between various groups within the same chain; these hydrogen bonds cause the polypeptide backbone to twist into a spiral shape. An important type of polypeptide helix is the α-helix.

The tertiary structure of proteins is often highly irregular. Almost no protein exists in a purely linear, helical form; rather, most proteins contain both helical and non-helical regions. In some proteins, there are virtually no helical regions at all. There are several stereochemical reasons why regions with the α-helical configuration, as well as other regular structural configurations, are relatively rare, despite the almost regular arrangement of hydrogen bonds in the protein backbone. One reason is the inability of proline (which lacks an amino group) to form hydrogen bonds; consequently, the regular arrangement of hydrogen bonds is disrupted in areas containing prolines. Another factor is the formation of disulfide bridges (S–S) between cysteine residues. When these cysteine residues are part of the same polypeptide chain, the helical structure is naturally disrupted. Perhaps the most important reason for the presence of irregularly structured regions in proteins is the diverse nature of the various protein groups. Each protein group strives to form the most energetically favorable secondary bonds with other groups. For example, a free hydroxyl group on tyrosine seeks a position that facilitates the formation of a hydrogen bond; the significant energy associated with such a bond would be wasted if the group were located near a hydrophobic side chain, such as that of isoleucine [176].

The phenomenon of gene overlap was first established experimentally in 1977 by the English biochemist F. Sanger. In the laboratory of molecular biology at the University of Cambridge, F. Sanger and his colleagues sequenced the nucleotide sequence of the genomic circular, single-stranded DNA of the phage φX174, which consists of 5386 nucleotides. The genome of this phage contains nine genes. A comparative analysis revealed that the number of amino acid residues in the proteins encoded by these genes exceeds the number of nucleotides in the DNA. This discovery led to the identification of an important molecular-genetic phenomenon known as "gene overlap" or "genes within genes." In the diagram showing the genome of phage φX174 in its unfolded form (Figure 8.2), gene B is located within gene A, and gene E is located within gene D. This arrangement increases the number of reading frames and, ultimately, compensates for the shortage of nucleotides in the DNA [95].

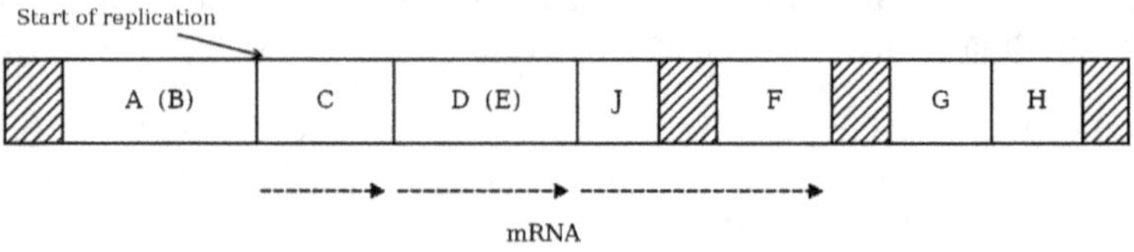

Figure 8.2. Schematic representation of genes. The genes are denoted by letters. Gene B is located within gene A, and gene E is located within gene D

At the M.V. Keldysh Institute of Applied Mathematics of the Russian Academy of Sciences, N.N. Kozlov conducted a series of studies aimed at mathematically analyzing the relationship between the structure of the genetic code and unusual methods of encoding genetic information—so-called "overlapping genes." The paper [20] investigates organisms that utilize a form of gene reading involving "shifts" in the genetic sequence; these organisms make use of overlapping genes, such as the phages φX174 and G4, as well as certain viruses. Through the mechanism of gene overlap, these organisms are able to encode their genomes in a more compact manner. "Overlapping genes refer to situations where, when a gene is read with the appropriate shift, none of the stop codons can be generated." The author introduces the concept of "elementary overlap" for individual amino acids. Gene overlap is considered complete if it occurs for any sequence of any length, regardless of the specific combination of amino acids. In other words, gene overlap means that there are no prohibited pairs of amino acids within the gene.

It has been found that overlapping genes are of utmost importance for identifying and analyzing new properties of the genetic code, which can only be obtained through mathematical methods. Overlapping genes are a key factor in reducing the amount of DNA that needs to be encoded.

During the analysis of large, decoded genomes, numerous genetic overlaps were identified; in the human genome, there were approximately 1700 of such overlaps. Although this data has already been published in several articles, certain aspects of this issue remain unclear. For example, what are the common locations of these genetic overlaps? What are the principles that govern their behavior? What approach can be used to handle gene overlaps that differ by only one single position? And how many changes are potentially permissible within such overlaps?

The relationship between the genetic overlap sequence and the two protein sequences is quite strict. For the longest overlaps, nucleotide substitutions that correspond to silent mutations are allowed. At each position where nucleotide substitutions are possible, pairs of codon synonyms can be used. Special attention is paid to the non-standard amino acids: Serine (Ser), Leucine (Leu), and Arginine (Arg). The irregularities in the coding of these amino acids provide additional possibilities for genetic overlaps that are not present in codes without such irregularities. Furthermore, cases of unusual genetic sequences have been considered; experts interpret these as rare instances of gene overlap, corresponding to the parallel reading of genes from different DNA strands.

It has been observed that the substitution of synonymous codons affects the functions of proteins in humans; some of these substitutions lead to changes in gene expression. Such cases underscore the importance of studying both potential and actual sites of silent mutations in the overlapping genes [20].

P.P. Gariaev [177] notes that the model of genetic code developed by F. Crick does not take into account the premises that it itself postulated. These provisions (the so-called Wobble hypothesis) are necessary for understanding the rules by which nucleotides pair during the process of reading codons with anticodons, as part of protein biosynthesis. From the rules established by Wobble's hypothesis, one fundamental characteristic of the genetic code emerges (an aspect that was previously overlooked): its homonymy. This is the second eigenvector of the degeneracy of the triplet code. The synonymy, as the first manifestation of degeneracy (i.e., the loss of code uniqueness among a set of distinct codons), was immediately identified. Homonymy refers to the codonic ambiguity regarding the first two nucleotides in codons. The third nucleotides can be any of the four possible types; therefore, they do not participate in the coding of amino acids. In other words, the ribosome reads the messenger RNA (mRNA) following a "two-out-of-three" rule. This was discovered in 1978 by the Swedish researcher U. Lagerkvist, but it has been ignored by the scientific community. It is absolutely clear that when the ribosome uses this rule, ambiguity arises in the reading of homonymous codons. For this reason, there is a risk of selecting the wrong amino acid or stop codon, which could lead to the synthesis of incorrect proteins and ultimately the death of the organism. However, the synthesis of protein is a very precise process. The ribosome never makes a mistake. Experiments indicate that the cell's protein-synthesizing machinery utilizes a linguistic-like mechanism of specific orientations when the ribosome accurately reads homonymous codons. According to Gariaev, the code's homonymity is a factor that enables the ribosomal apparatus, and thus cells as a whole, to function at a level of quasi-intelligence. P.P. Gariaev's findings are of great significance not only from a theoretical-biological perspective but also as a strategic guide for the future development of genetics and biology as a whole.

In order to properly design the model, it is obviously necessary to understand not only the chemical processes but also the physicochemical phenomena that occur within the cells of an organism.

In the 1970s, the concept of "molecular machines" emerged and became widely accepted; it has since firmly established itself in the scientific literature. A molecular machine is a macromolecular structure, usually of natural origin (for example, a protein molecule or a protein complex). It performs functions such as converting chemical energy into energy that drives the directed movement of macromolecules or individual fragments, as well as transporting molecules and ions across biological membranes. The idea of enzymes as nanoscale molecular machines implies that they possess a relatively small number of distinct (mechanical) degrees of freedom—degrees of freedom related to the coordinated movement of atomic groups or the presence of movable parts within the molecule. This means that individual components of a protein or protein complex can be viewed as elements of a macromolecular structure, analogous to the parts of macroscopic machines and mechanisms.

From a physical perspective, the main characteristic of molecular machines and mechanisms (MMMs) is the combination of mechanical and statistical properties within them.

The functioning of a living cell and all multicellular organisms is primarily driven by free energy (Adenosine Triphosphate, ATP). The vast majority of ATP molecules are produced by proton-driven ATP synthases, which are the smallest nanoscale molecular motors found in nature.

The mechanical properties of MMM are determined by large-scale structural rearrangements of macromolecules, resulting from the movement of relatively large fragments of protein molecules along a small number of specific degrees of freedom.

The statistical properties of polymers (MMMs) are determined by the thermal movements of individual atoms and small groups of atoms within the macromolecules.

One of the most active areas in biophysics relates to the study of rotating molecular motors (MMMs).

Biological machines (BMMs) whose operation is based on the directed rotation of their rotors are typically integrated into biological membranes. They are activated by the electrochemical potential differences generated within these membranes. To date, two families of naturally occurring nanoscale rotary "electric motors" have been identified: bacterial motors and the $F0F_1$ complex, which is an ATPase protein complex responsible for the synthesis of adenosine triphosphate (ATP).

Bacterial electric motors consist of a stator embedded within the membrane of the bacterial cell, as well as a rotating rotor. The source of energy for the rotation of the rotor is the transmembrane electrical potential difference. This complex nanoscale device features a specialized molecular "transmission mechanism" that allows the direction of the rotor's rotation to be changed. Additionally, the bacterial cell contains a molecular "processor" analogous to a modern computer, which controls the behavior of the cell and its organization in space. This macroscopic-like "processor" receives, stores, and processes information about the chemical composition of the surrounding environment or levels of illumination, obtained from specialized receptors located on the cell surface. It then generates appropriate signals to switch the direction of the rotor's rotation. Thanks to this mechanism, bacteria can change the direction of their movement and navigate towards a target location, or conversely, move away from an environmentally unfavorable area [178]. This research discusses current understanding of the physicochemical mechanisms underlying proton ATP synthase—the primary macroscopic enzyme responsible for catalyzing the formation of ATP molecules.

$F0F_1$– ATP synthase is one of the most important molecular complexes in a living cell. It serves as a unique energy converter within the cell and contains two rotating motors: F_0 and F_1.

The ATP synthase complex, embedded in the cell membrane, functions as a macroscopic molecular machine responsible for energy conversion. It is activated by the transmembrane electrochemical potential difference of hydrogen ions ($\Delta\mu H+$). The electric current flowing through the F_0 membrane complex (protons serve as the charge carriers) drives the rotation of a rotor (an embedded drum within the membrane, denoted as s_n). This molecular "electric motor"-type structure is connected to the rotor of the mechanistic motor F_1 (the γ subunit); the rotation of the F_1 rotor facilitates the synthesis of ATP from ADP and P_i.

$F0F_1$ – ATP synthase is a reversible molecular machine capable of catalyzing both the synthesis and hydrolysis of ATP. As the rotor of the F_1 motor rotates in a direction opposite to that of the rotor of the electric motor F_0, $F0F_1$-ATP synthase functions as a proton pump, generating a $\Delta\mu n+$ gradient. It is important to note that $F0F_1$-ATP synthase is just one of the many essential molecular machines found in living cells; hundreds of billions of such macromolecular motors are active within every organism, facilitating the production of ATP – the "chemical fuel" essential for cellular metabolism.

Within a single day, the human body synthesizes more than 50 kilograms of ATP molecules, which are then utilized by the body to perform various types of chemical, osmotic, and mechanical processes. There is now compelling evidence indicating that the rotor of the $F0F_1$-ATP synthase complex rotates during its operation. The rotation occurs in discrete steps, with each step corresponding to a 120° change in orientation. The average rotational speed of the rotor increases as the concentration of ATP in the solution surrounding the enzyme rises, as well as when the length of the molecular "marker" used to observe the rotation is decreased; this can result in a rotation rate of approximately 10 revolutions per second. The mechanism of $F0F_1$-ATP synthase falls within a class of mathematical models known as "mesoscopic models" or Markov-Fokker-Planck (MFP) models. These models represent an intermediate approach between those that use molecular dynamics to simulate the motion of individual atoms and those that describe the transitions between a relatively small number of discrete chemical states.

In the work [179], using the example of studies on electron transport processes in photosynthesis and respiration, the main functions of the mathematical model are demonstrated: its use as a tool for testing theories and hypotheses formulated based on experimental data; as well as theoretical analysis of the nature of complex experimental relationships that cannot be simply explained qualitatively. It is also possible to predict new properties of biological entities based on the model, even if these properties do not directly emerge from the experiments. When constructing mathematical models of electron transport processes occurring within macromolecular complexes, the authors start from probabilistic assumptions.

In recent years, many researchers have begun to acknowledge the advantages of quantum computers over classical computers. The need for a quantum computer arises when scientists attempt to study complex biological systems using the methods of physics. A quantum computer uses processes that are inherent to the quantum realm, rather than ordinary classical algorithms, for performing calculations. Genetic molecules exist precisely on the basis of such principles. Quantum computers will enable us to calculate the structure and behavior of quantum systems, including protein molecules. If science is able to implement quantum algorithms in a real device, it will be possible to compute the function f(x) for various values of x simultaneously, resulting in a significant saving of computing time. The efficiency of quantum computers is related to quantum parallelism—a unique property of quantum computing. [167]

One of the characteristics of genetic information is mutation. We will discuss this phenomenon in more detail later on.

8.1. MUTATIONS AND THEIR CHARACTERISTICS

In a protein-based text, errors in the amino acid sequence can occur. Such errors alter the biological function of the protein and have very serious consequences for the organism. Genetic diseases are extremely rare, which once again demonstrates the nature's remarkable ability to protect itself quite effectively against hereditary disorders related to metabolic processes. Nevertheless, when such disorders do occur, they are usually associated with changes in the protein structure or a loss of enzyme activity. For example, the substitution of glutamic acid (Glu) at the sixth position of the β-chain of human hemoglobin with valine (Val) leads to the severe disease sickle cell anemia. Sickle cell anemia is a hereditary blood disorder; in affected individuals, red blood cells take on a sickle-shaped appearance. During the first months of the patient's life, this condition can result in blockages of small blood vessels, bleeding, and heart attacks. Hemoglobin A, the main type of hemoglobin in adults, consists of four polypeptide chains—two identical α-chains and two identical β-chains (2α2β). It should be noted that this has nothing to do with the α- and β-helical secondary structures (α-helix and β-sheet) of proteins. In his work [20], N.N. Kozlov points out that both normal and sickle cell hemoglobins contain the same types of chains; however, the specific amino acids at certain positions in these chains differ. In normal hemoglobin, a glutamic-acid residue is present at the sixth position of the β-chain, whereas in sickle-cell hemoglobin it is replaced by a valine residue.

	1	2	3	4	5	6
Normal	Val	His	Leu	Thr	Pro	Glu
Sickle cell anemia	Val	His	Leu	Thr	Pro	Val

In this case, the differences between the normal and mutant versions are due to a single amino acid substitution in the respective protein. One of the challenges in medical treatment for this disease is to learn how to correct gene defects.

Proteins play an invaluable role in protecting the body from diseases. Lysozyme is an enzyme that is found in tears and protects the eyes from a variety of microorganisms. The protection of the organism must be aided by medicine.

The more the composition of a drug resembles the substances found in the human body, the more effective its effect will be. Lately, the mysterious word "peptides" has increasingly appeared on the packaging of pharmaceutical products. Some experts refer to medications made based on peptides as the "XXI century medicine." Peptides, as we have already mentioned, consist of a certain number of amino acids that are connected to each other by so-called peptide bonds. If a chain of amino acids is short, it is a peptide; if it is long (more than 50–100 amino acids), then it is a protein. The journal "Itogi" contains a quote from the deputy director of the Institute of Molecular Genetics at the Russian Academy of Sciences, Academician N. Myasoedov: "Peptides are involved in the formation of various physiological functions—ranging from simple to complex ones, even including behavioral aspects; Scientists have long been pondering whether it is possible to repair damaged regions of protein molecules using medications based on peptides [180]. A group of genetic scientists proposed the hypothesis that age-related changes lead to alterations in the primary, secondary, and tertiary structures of DNA, RNA, proteins, enzymes, and hormones, all of which are involved in the functioning of an organism. During the aging process, errors accumulate in the organism, causing the DNA to begin "printing" distorted information. The young organism contains more enzymes that are capable of repairing "errors" in the DNA. However, as one ages, this ability declines or even disappears altogether. In 1980 Schmukler-Reiss and S. Goldstein discovered that a particular segment of human DNA, which consists of 340 nucleotides, occurs 100,000 times within a cell. Another segment, of 680 nucleotides, is repeated 35,000 times. As cells age, repetitions become increasingly rare. It appears that this is not a coincidence. The DNA molecule in the cell nucleus is twisted into a superhelix structure. The more tightly the helix is coiled, the greater the number of turns it contains. It turns out that the younger the cell, the more tightly the DNA is wound into a spiral. In the authors' opinion, the reduction in the number of helical turns during aging is accompanied by a decrease in the synthesis of proteins that regulate cell division [181].

Mutations (from the Latin mutatio – change) are natural or artificially induced alterations in an organism's hereditary traits, resulting from rearrangements and disruptions in the genetic material (chromosomes and genes). Mutations are the basis for genetic variation in living organisms. The founder of the theory of mutation was the Dutch scientist H. De Vries, one of the creators of this scientific framework. The theory of mutation emerged following the discovery of hereditary changes in an organism's traits and characteristics. Human populations always carry a certain amount of harmful mutations; this is what is known as the "genetic burden" or the inherited load on humanity. The extent of this genetic burden depends on the frequency of mutations and the intensity of natural selection.

According to the theory of mutation (Russian scientist S.I. Korzhinsky, 1899; Dutch scientist H. De Vries, 1901–1903), sudden, abrupt mutations are the decisive factor in evolution, immediately leading to the emergence of new species. Natural selection was considered to play a secondary role. In the subsequent integration of genetics and Darwinism (1920–1930), it was established that evolution can only occur through the natural selection of mutations [46]. Mutations represent discrete changes in the various units of the genetic code that govern the inheritance of information in living organisms

(Timofeev-Resovsky, Vorontsov, Yablokov, 1969). These abrupt changes in genetic structures occur spontaneously and are transmitted to subsequent generations according to Mendelian laws.

De Vries, in his work "The Theory of Mutation," formulated the main principles of this theory, which have not lost their significance to this day.

1) Mutations occur suddenly, without any transitional phases.
2) The new forms are completely constant (stable) in their manifestation/expression.
3) Mutations, unlike non-hereditary changes (fluctuations), do not form continuous sequences, nor do they cluster around a mean type (mode).
4) They go in different directions; they may be useful, harmful, or neutral.
5) The detection of mutations depends on the number of individuals that have been analyzed.
6) The same mutations can occur again.

H. De Vries worked with the results of mutations, specifically their phenotypic manifestations. Modern theories of the mutation process focus on the nature of mutations and the mechanisms by which they occur.

Gene mutations refer to the direct alteration of genes that are located within chromosomes, as a result of external factors.

As Alberts noted: "Spontaneous DNA damages that remain unrepaired rapidly alter the DNA sequence." [182]

Even collisions with other molecules due to thermal motion cause significant changes. On any given day, the DNA in each human cell loses approximately 500 purine bases as a result of temperature-induced damage; these and other random DNA alterations are removed by the cell's repair mechanisms.

The factors that cause mutations (mutagens) include radiation, toxic chemical compounds, and viruses.

As a result of mutations, new alleles of the corresponding genes are formed, which is the basis for the genetic diversity of living organisms. Alleles differ from each other in the composition of nucleotides within the organic regions of DNA.

The largest and most significant portion of all mutations consists of gene mutations. These can be divided into two categories. The first category includes mutations that are related to the replacement of nitrogenous bases; such mutations account for no more than 20% of spontaneous mutations. The second category, which comprises the majority of mutations, is caused by shifts in the reading frame of the genetic material, and includes insertions and deletions (the removal of DNA segments).

Gene mutations refer to changes in the number or sequence of nucleotides within the structure of DNA, specifically within a particular gene. These alterations result in changes in the quantity or quality of the protein products synthesized by that gene.

There are two types of base replacements:

- The replacement of one purine with another purine, or of one pyrimidine with another pyrimidine, is also allowed.
- This involves replacing purine with pyrimidine, or vice versa.

A mutation that results in the substitution of an amino acid can be "conservative." For example, a point mutation that replaces the codon CUU with AUU will lead to the replacement of leucine with another neutral, non-polar amino acid: isoleucine. In such a case, the function of the protein may be partially preserved.

Not all base substitutions are manifested as mutations. Due to the redundancy of the genetic code, all amino acids except methionine and tryptophan are each encoded by two, three, or more codons.

Sometimes, the substitution of a nitrogenous base can result in one codon being converted into another codon that codes for the same amino acid in the protein molecule.

Nonsense mutations are usually lethal. They reduce the activity of all structural genes within the same operon.

A significant portion of all spontaneous mutations consists of mutations that involve shifts in the reading frame of the DNA. Most of these mutations have been identified in sequences that consist of identical nitrogenous bases [183].

When examining a very long text that has been broken down into short words of equal length (codons), it is possible to determine the effect of inserting or deleting characters.

In the case of single (+) or (-) mutations, that is, when one symbol is added or removed, the entire correct phase of translation is disrupted. For mutants with the (++) and (--) mutations, the correct translation phase is not restored; however, for mutants with the (+ -) and (- +) mutations, the correct translation phase should be restored.

If the correct translation phase can be restored in triple mutants (+++) and (---), it means that the codons are triple in length (i.e., they consist of three nucleotides each). This is because the phase shift observed is also triple in magnitude. If there are non-coding triplet sequences (nonsense sequences) within the genetic code, then the phase shift will result in a new, random sequence of codons being generated; among these new codons, there may still be nonsense sequences.

Nonsenses may not occur in small regions of the genetic code, but they will definitely appear in larger regions and become obstacles to translation. If the codon system contains 64 triplet codons, and the number of nonsense codons is two to three, then on average they should occur over shift regions larger than 20–30 codons (40–60 nucleotide sequences, n.p.). Therefore, in many double mutants of the (+-) and (-+) type, the translation phase should be restored over distances of no more than 40–60 n.p., when nonsense barriers are practically rare. This has been confirmed experimentally. Therefore, the genetic code is highly degenerate and contains several non-coding triplets — nonsenses [102].

Different mutational substitutions have different effects on the structure of a protein. Mutations that significantly alter the hydrophobicity of an amino acid residue should have a greater impact on the protein's biological properties than those that only slightly modify its hydrophobicity. The first type of mutation is more detrimental to the survival of a species; therefore, the genetic code has been designed in such a way as to give a selective advantage to mutations of the first type. It is necessary to identify these characteristics of the genetic code.

M.V. Volkenstein discussed the second possibility in his work "Molecular Biophysics" [184]. He presented a dictionary of codons, denoted by xyz, in the form of a square table (Table 8.2). In Table 8.2, hydrophobic residues are located in the dark cells. Notably, when the second position of the codon is U (y = U), the residue is always hydrophobic. The computational data he obtained agree with the code table (Table 8.2). Replacing the middle position y of the codon poses the greatest risk, as it leads to the most significant changes in the protein's hydrophobicity; replacing position z, on the other hand, is the least dangerous. This pattern reflects the physical significance of the genetic code. The advantage of "correct" mutations is determined by the three-dimensional structure of the protein in an aqueous environment, as well as by the specific physical properties of water. Thus, the correspondence between codons and amino acids is governed by the physics of water. The genetic code exhibits a high degree of reliability—although not absolute—in distinguishing between "correct" and "incorrect" mutations.

M. Ichas indicates that both too high and too low mutation rates are evolutionarily disadvantageous; there is a need for an intermediate, optimal frequency [185].

Table 8.2

y x	A	C	G	U	z
A	Lys	Thr	Arg	Ile	A
A	Asn	Thr	Ser	Ile	C
A	Lys	Thr	Arg	Met	G
A	Asn	Thr	Ser	Ile	U
C	Gln	Pro	Arg	Leu	A
C	His	Pro	Arg	Leu	C
C	Gln	Pro	Arg	Leu	G
C	His	Pro	Arg	Leu	U
G	Glu	Ala	Gly	Val	A
G	Asp	Ala	Gly	Val	C
G	Glu	Ala	Gly	Val	G
G	Asp	Ala	Gly	Val	U
U	Stop	Ser	Stop	Leu	A
U	Tyr	Ser	Cys	Phe	C
U	Stop	Ser	Trp	Leu	G
U	Tyr	Ser	Cys	Phe	U

The human genetic system is structured like a piece of text: it has its own "letters," "words," punctuation marks, and rules of spelling. A person receives a randomly selected half of the genes from each parent. There are genes in the human body that are inherited strictly along one lineage, without any mixing. For example, the Y chromosome is present only in males and is passed from the father; since it has no counterpart (i.e., no matching chromosome), the genetic information contained within it is transmitted from generation to generation without any changes, except for rare and random "errors" (mutations). Since the Y chromosome is only passed from fathers to sons, mitochondrial DNA (mtDNA) is received in its unaltered form from the mother by both sons and daughters. However, only daughters pass on this mtDNA to their offspring.

By the late 20th century, the successes of molecular biology enabled scientists to directly read genetic sequences, identifying even single-nucleotide discrepancies. However, each research laboratory decided for itself how to conduct genetic analyses; as a result, the methods used varied from lab to lab, making it impossible to compare the results of their studies. Later on, certain standards for such research began to emerge. It became necessary to develop a common framework that would allow for the standardized evaluation of the quality of conducted experiments using consistent methodologies.

Different tests are based on different types of DNA. The DNA in the cell nucleus is passed from parents to children. Nearly 99.9% of this DNA sequence is the same for all humans, but 0.1% varies in certain parts of the genome—where a T is replaced by a C, or a G is replaced by a C. These differences (referred to as mutations) account for the incredible diversity among humans. Most mutations are single-nucleotide polymorphisms (SNPs). In general, these are harmless changes that occur spontaneously and are passed down from generation to generation.

Pairwise comparisons allow for the rapid construction of a branching diagram in which each branching point represents a mutation in a specific nucleotide, and each branch corresponds to a carrier of one of the variants (either the mutant or the original version). Such branches in genetics are known as haplogroups.

The integration of mathematical methods into biology, including biometrics and statistics as specific branches of modern mathematics, reflects the need to intensify our understanding of biological phenomena.

The fundamental theoretical principles on which biometric analysis methods are based rely on mathematical statistics, probability theory, and the law of large numbers. These principles help to identify patterns in the behavior of random events within large datasets.

To avoid obtaining inaccurate values, it is necessary to determine the sample size before collecting the primary data. The formula used to calculate the sample size is:

$$n = \frac{N}{N(\frac{d}{t}) + 1} \tag{8.1}$$

where n is the desired sample size.

N – the size of the general population;

t represents the reliability criterion, which corresponds to the selected level of probability: $t_0.95 = 2.0$; $t_0.99 = 2.6$; $t_0.999 = 3.29$.

d – the allowable difference between x̄vyb and x̄gener.

where x̄sample and x̄population represent the arithmetic averages.

Most of the non-coding, redundant nucleotide sequences in the genomic DNA of eukaryotes do not contain any genetically important information. Eukaryotic cells possess mechanisms that provide additional protection for the genomic DNA against mutation. Analysis of the genomes of modern eukaryotes reveals that the evolutionary changes that led to the incorporation of these redundant nucleotide sequences were accompanied by significant genetic adaptations regarding the stability of genetic information. In particular, the much higher abundance of redundant nucleotide sequences compared to coding sequences must inevitably result in a corresponding decrease in the likelihood of mutations occurring in the coding regions and other functionally important parts of the genome, due to the influence of both endogenous and exogenous mutagens.

The effect of excess nucleotide sequences on the number of mutations arising from replication errors in the coding sequences of the genome is explained in work [186] as follows. It is assumed that the length of the original genome, which does not contain excess nucleotide sequences, is N base pairs. As a result of replication errors, the genome contains an average of a mutations, where a is independent and random. During evolutionary processes, excess nucleotide sequences are incorporated into the genome; their total length is nN base pairs. The total length of the modified genome becomes (n + 1)N base pairs. The overall number of mutations that occur on average in the modified genome due to this mechanism must increase by a factor of (n + 1), resulting in a value of a(n + 1). The probability of a single, independent, and random mutation occurring in a non-coding region of the genome (P(1)) is proportional to the length of that region.

$$P(1) = \frac{nN}{(n+1)N} = \frac{n}{n+1}. \quad (8.2)$$

At the same time, the probability of all a(n + 1) mutations occurring in the redundant parts of the genome will be equal to:

$$P[a(n+1)] = \left(\frac{n}{n+1}\right)^{a(n+1)}, \quad (8.3)$$

This is because the probability that all of the a(n + 1) independent events will occur simultaneously is equal to the product of the probabilities of each event occurring individually.

When a = 1 (that is, when, on average, one mutation occurs during the replication of the original genome), and n is a sufficiently large value, this expression tends to approach e^{-1}, which is approximately equal to 0.36. Therefore, in this case where n = 100 (which roughly corresponds to the ratio of non-coding to coding nucleotide sequences in the genomes of humans and other mammals), the probability that none of the mutations occurring in the hypothetical transformed genome due to replication errors will occur in its coding regions will be quite high, amounting to approximately 0.36. This means that, on average, every third daughter somatic or germ cell arising from the replication of a hypothetical genome, which contains a sufficient amount of non-coding, redundant nucleotide sequences, will be completely free from mutations that occur in the coding regions of its genome through this mechanism.

As the number of mutations in the precursor genome increases (i.e., when a >> 1), the likelihood of all mutations occurring in the non-coding regions of the genome decreases rapidly. However, since these mutations will be distributed across both coding and non-coding regions in proportion to the length of each region, the total number of mutations in the coding portion of the genome remains unchanged. Therefore, the evolutionary incorporation of a large number of non-coding nucleotide sequences into the precursor genome does not increase the number of mutations that occur in the coding regions due to replication errors. The situation regarding mutation formation in a hypothetical genome under the influence of exogenous and endogenous mutagens is fundamentally different from the one discussed here. If it is assumed that the evolutionary transformation of the genome, which resulted in the incorporation of n non-coding nucleotide sequences, was not accompanied by an increase in the number of intranuclear mutagens, then the genetic consequences of such a transformation would be much more significant. Similarly to the previous case, the probability of a particular mutagen (M_k) being inserted into a non-coding region of the hypothetical genome is equal to…

$$\frac{n}{n+1}. \tag{8.4}$$

The probability that all k mutagens will end up in the non-coding regions of the new genome P(k) is equal to:

$$P(k) = (\frac{n}{n+1})^k . \tag{8.5}$$

As n becomes large and k becomes small, the value of P(k) approaches 1, which indicates that the event in question is nearly certain to occur. In other words, the greater the proportion of non-coding nucleotide sequences in the genomic DNA, the more likely it is that all intranuclear mutagens will end up within these non-coding regions. For a hypothetical genome with n = 100, the probability of a single mutagen landing in the coding region is equal to 0.01, which is quite low. As a result, the total number of mutagens that will interact with the coding nucleotide sequences will be reduced by 100 times. Moreover, there will be 100 times greater protection of the coding regions of the evolutionarily transformed genome against mutations caused by intranuclear mutagens compared to the original genomic precursor. As k = n + 1, as n increases, the function P(k) will approach e^{-1} (which is approximately equal to 0.36), and it will decrease rapidly as k continues to increase. However, this refers to the probability of fully protecting the coding sequences of nucleotides.

The relative protection provided by a genome will be equal to the proportion of chemical mutagens in the total pool of mutagens that interact with the coding sequences of nucleotides. This protection will be inversely proportional to the total length of the redundant (non-coding) sequences in the transformed genome, i.e., inversely proportional to "n". This conclusion only applies to mutagens that are directly generated within the nucleus, for example as a result of the interaction of ionizing radiation with nuclear materials. However, it is likely that the majority of mutagens enter the nucleus from the cytoplasm via radial diffusion through the nuclear membranes. One factor limiting the penetration of mutagens from the cytoplasm into the nucleus is their surface area; indeed, the probability of contact between cytoplasmic mutagens and the nuclear membrane is directly proportional to the surface area of the nucleus.

Each person carries 5 to 7 genetic mutations, which, under fortunate circumstances, may never be expressed at all.

Genetic scientists have proven that if a mutation is present in both spouses, the probability of having a child with the disease is 25%.

Advances in genetics over the past few years have shown that DNA studies can not only identify existing diseases in humans, but also predict future illnesses in unborn children. As of 2005, genes responsible for approximately 500 hereditary diseases had been identified. In Russia, practicing geneticists identify people's predispositions to 60 different diseases. To date, approximately 20 hereditary diseases have been identified, which are caused by dynamic mutations. The etiology of these diseases is associated with the expansion of trinucleotide repeats (CGG)n, (CTG)n, and (CAG)n. The cause of the damaging effect in some cases of dynamic mutations is the loss of cellular function due to the blockade of gene expression; in other cases, it is the presence of protein products with abnormal structures that have a detrimental impact on cell viability. Mutations of the type described are similar to other types of expansion mutations: for example, the expansion of pentanucleotide repeats (specifically, the ATTCT repeat sequence in a non-coding region of the gene). In the wild type, there are approximately 20 copies of this repeat; in the mutant allele, the number of copies can range from 800 to 4500. The person carrying the mutation does not know whether they will develop the associated disease, and the doctor can only inform them about the risk and explain what might happen if the disease does occur. The mother of the future child, if she is a carrier of a dangerous mutation, can have the fetus examined in a timely manner. Prenatal DNA testing in the early stages of pregnancy can detect whether there are mutations in the fetus; as a result, the birth of a child with an incurable disease can be prevented.

For example, according to Natalia Irinina, the head of the medical genetics consultation service in Vladimir, most patients follow the doctors' advice and prefer not to take risks; however, some do disregard their recommendations. There have been cases where couples refused to terminate their pregnancies but later changed their minds and decided to keep their sick children in the hospital [2].

To make a diagnosis, geneticists often need to examine the patient's blood relatives. However, it is not always possible to obtain their consent due to a lack of information. Currently, in Europe, according to data from the European Commission for Scientific Research, approximately 700,000 genetic tests are conducted each year – which is hundreds of times more than ten years ago. In Russia, such precise statistics are not yet available. Nevertheless, there are around 100 medical-genetic consultation centers throughout the country (almost in every large city), and these centers receive an average of several dozen patients per day [2].

Often, treatment is aimed at alleviating the symptoms rather than addressing the genetic root cause of the disease. Nevertheless, medical professionals are constantly finding ways to combat diseases in their early stages.

Total loss of vision (Leber's amaurosis) is a disease caused by a defect in the RPE65 gene. As a result of this gene mutation, the light-sensitive cells in the retina die and do not regenerate, leading to the gradual loss of vision. British medical researchers have found a way to combat this disease; they have restored vision to one of their patients with Leber's amaurosis. A virus carrying a healthy gene was injected into the patient's retina using a fine needle. The experiment was successful.

Numerous discussions regarding the criminality associated with biological factors, both in Western and domestic literature, focus on the issue of chromosomal abnormalities [187]. Two specific chromosomal abnormalities have been identified as potentially criminally significant: Klinefelter's syndrome (a sex chromosome disorder in males with the karyotype 47/XXY) and "Y-disomy" syndrome (another sex chromosome disorder in males with the karyotype 47/XYY). It is important to note that the normal sex chromosome configuration for males is 46/XY.

Klinefelter's syndrome is quite often accompanied by severe mental disorders. It has been observed that the higher the number of X chromosomes in the genotype, the more pronounced the intellectual disability—sometimes even reaching the level of idiocy.

The additional Y-chromosome, first discovered by A.A. Sandberg in 1961, was studied in 197 criminals with mental disabilities (the study was conducted by R.A. Jacobs). These cases were considered to be related to the criminal tendencies and aggression of the individuals possessing this chromosome. However, many believe that this conclusion is fundamentally incorrect. It equates mentally ill and therefore incapacitated persons with criminals, and it links behaviors "that do not conform to our notions of the psychology of a normal person" with crimes—something that does not align with the principles of criminal law regarding guilt and responsibility.

The behavior of a mentally ill person that poses a danger to society is not a crime; the principles governing such behavior are psychiatric in nature, rather than criminal.

The incidence of chromosomal abnormalities remains relatively constant over time and is statistically independent of a country's socio-political system. For the 47/XYY aberration, the frequency is 1 in 1000 births; for Klinefelter syndrome (47/XXY), it is 1 in 1400 newborn boys [188].

The rate of change in criminal behavior is completely incomparable to the pace of human genetic evolution. The genotype of modern humans emerged approximately 40,000 years ago. Of course, this does not mean that natural selection has come to an end; genetic development is an endless process, albeit one that occurs very slowly. Criminal behavior, on the other hand, changes within just a few years or decades. Modern genetics has established that acquired characteristics—whether physiological or, even more so, moral traits of personality—are not inheritable [189].

DNA analysis refers to methods used for studying biological entities. For instance, the results of DNA analysis can be used to identify a person based on the biological traces left at the scene of a crime. The presence of bacteria on a person's skin depends on various factors: medical conditions, genetic factors, the use of medications (such as antibiotics), living environment, and so on. This information provides insights into both individual and group characteristics of that person. In biology, all living organisms are generally divided into two main categories: prokaryotes (organisms without a nucleus) and eukaryotes (organisms with a well-defined nucleus within their cells). Therefore, the current classification of methods for studying biological entities can be understood in conjunction with the modern classification of biological organisms themselves.

David Goldstein, a geneticist at the University of London, predicts that doctors will soon start prescribing medications based on a patient's genetic profile. "To make use of genetic research findings for the personalized selection of drugs, it is crucial that the public has trust in how genetic information is being used. If we overuse genetics to inform people about their ancestry, it will lead to skepticism regarding the reliability of information relating to their drug responses," he says.

It is well known that genetics, as a scientific discipline, as well as its practitioners, suffered severe setbacks and persecution during the 1940s and 1950s. Many scientists, including mathematicians and physicists, participated in the revival of genetics. In modern biology, genetics occupies a special position; it is aptly referred to as the "philosophy of biology." Today, genetics has become a unified and comprehensive science that employs both biological and physicochemical methods to address some of the most significant biological challenges.

In the advanced capitalist countries, primarily the United States, Japan, and West Germany, there is a clear understanding of the priority of fundamental genetic and molecular-genetic research. As a result, genetics has developed at a rapid pace in these nations. In Russia, genetics declined during the period of the cult of personality; during the period of stagnation, it progressed extremely slowly.

According to medical associations in Western Europe and the United States, 15% of people require genetic medical assistance. In countries with high living standards, genetic clinical research has become part of modern healthcare. In Russia, however, access to genetic medical services for the general population is limited due to organizational and financial reasons. Nevertheless, genetic research activities are still underway in the country. The Genetics Laboratory at the Kurchatov NBICS Centre (Nano-, Bio-, Information, Cognitive and Social Technologies) is equipped with a complete set of facilities and equipment for genome sequencing. More than 20 genomes have already been sequenced at this Russian scientific institution [190].

The biggest problem with organizing research in genetics is its practical lack of transparency for those who are willing to invest money and develop new technologies in this field. In the developed countries of the world, on the other hand, a transparent scientific environment has been in place for several decades, with its own investors.

The scientist must come up with a scientific idea, secure a grant for that idea, and present results that will serve as a basis for investors to continue investing money in the project. Science can only advance through transparency. The world of science is based on markets and reputation.

Some articles in scientific journals are essentially repetitions of previous work, with the material just being presented in a different way. Such articles cannot possibly be "productive" or lead to new insights, because if those ideas and findings went unnoticed before, they will likely become obsolete within 3–5 years. However, it is very difficult to predict in advance which specific scientist will present which results.

Some articles contain unfounded hypotheses that are accepted as truths. Some results seem unexpected only because they were actually incorrect. This undermines the authority of genetics as a science.

It is generally believed that the more unexpected a result is, the more critically it should be viewed, as there is a higher likelihood that the result is incorrect. Articles containing false data not only harm their authors but also lead to futile or hopeless projects.

Some scientists in the field of genetics identify two potential sources of errors. Firstly, they argue that the exceptionally large size of the genome means that unusual results may occur randomly during research and analysis. This is why it is necessary to employ theories of probability and mathematical statistics; to do so, one must have expertise in relevant research methods and analytical techniques. Secondly, all high-throughput genomic technologies are prone to producing errors, some of which may be systematic in nature and could initially appear to be interesting biological phenomena. As a result, researchers might draw incorrect conclusions [4].

It is recommended to establish and adhere to quality control standards for research results. Genomics has only just begun to make its way into biomedical studies, so rapid success should not be expected. At the same time, geneticists should not lose faith in the ultimate success of genomic medicine, which has the potential to transform healthcare.

In solving problems in the field of genetics, personal medicine is of particular interest. If a country is able to equip medical institutions with sequencers, then everyone will benefit.

Genetic illiteracy will be overcome in the same way that general illiteracy was combated in Russia at one time.

Mankind is suffering, and it needs to be treated. To do this, we must employ new, more efficient methods—especially those capable of accurately decoding any genetic information.

CONCLUSION

The content of this material introduces readers to the mathematical methods used for analyzing and recognizing genetic code as well as genetic information. The book explains in considerable detail a range of basic concepts and definitions, including: complex systems, information theory, recognition methods and their classification, inductive approaches, Bayesian procedures, Markov chains, intelligent and expert systems, difficult-to-solve problems, and the fundamentals of genetics, among others.

Today, complex systems are studied in many fields of science, including biology. This monograph discusses the concept of a complex system and provides the foundations for a systematic approach to the study and classification of such systems.

Currently, the information possesses a unique characteristic compared to other sources: it can be used repeatedly in various fields of science.

Recognition is one of the most versatile abilities associated with solving many vital tasks in life. Tasks related to classifying features, phenomena, and processes, as well as the actual recognition of those entities, are closely interconnected with each other.

Mathematics has successfully tackled the problems that life has posed to it. Probability theory, mathematical statistics, theory of differential equations, theory of optimal control, and computational mathematics have all been involved in solving practical problems and have effectively met human needs. However, the development of mathematics has not always been smooth; there have been periods of decline as well as periods of growth in mathematical thought. The 20th century witnessed a crisis in the foundations of mathematics. In recent decades, an increasing number of theorems in mathematics have proofs that are utterly beyond human comprehension without the aid of computers. Many mathematicians seriously question the reliability of such proofs, since they have never been fully verified by hand. As a result, pure mathematicians are extremely skeptical of proofs obtained using computers.

The crisis in mathematics was also fueled by the widespread appearance of overly difficult problems (problems that require excessive computational effort to solve).

The inadequacy of the methods and foundations of mathematical logic for solving combinatorial problems led to the need for exploring unconventional approaches, as well as logical tools and information representation formats and processing algorithms that differ from those of classical mathematical logic and its foundations.

Nevertheless, it is the combination of the complex internal structure of mathematics with real-world application problems that constitutes the driving force behind the advancement of this science.

Special contributions to the development of mathematical methods have come from the theory of probability.

Markov chains have once again seen significant development in science. In biology, Markov chains are used to study the patterns of development in various species of animals and plants.

The Bayesian procedure is adopted as the basis for the inductive approach to recognition. It should be noted that if one of the object classes is missing from the training dataset, then any recognition procedure will perform poorly. A key aspect of justifying the inductive reasoning-based recognition procedures is the presence of all classes in the dataset, as well as averaging the errors across multiple training datasets.

In an inductive approach, the concept of an "exact solution" does not exist. The use of approximate solutions has made it possible to develop efficient inductive procedures with polynomial error estimates, which allow for the computation of a solution in just one step [134].

Modern mathematics should be considered a combination of deductive and inductive approaches. The realm of deductive methods is focused on obtaining precise solutions to problems of small scale; the realm of inductive methods is focused on solving problems of large scale using approximate calculations (e.g., [149]).

A brief historical overview of the development of genetics is provided, as well as an explanation of the structure of DNA. The concept of "genetic code" is also discussed. Special attention is given to the statistical analysis of the genetic code and the role of mathematics in genetic research.

Attention has been paid to the logical methods used in recognition processes. The characteristics and challenges of genetic research and recognition techniques have been reflected in the analysis.

In the process of presenting this material, quotations from individual scientists are included; these quotations fundamentally change the previously established concepts in the field of genetics. For example, the non-coding regions of the genome that do not encode proteins (referred to as "junk") actually play an important regulatory role. A comparative analysis of the non-coding regions in the genomes of higher eukaryotes revealed a mosaic structure consisting of highly conserved and diverse parts. The so-called "junk" DNA also performs codogenic, genetic functions [Akifeev, 2004].

Scientists have observed that the substitution of synonymous codons also affects the functions of proteins in humans, leading to changes in the expression of the gene [20].

However, it should be noted that to date there have been no fundamental shifts in our understanding of the genetic code. In educational literature, we do not see any new advancements in the field of genetics; neither has any of the basic principles of official biology textbooks been revised.

It is still not clear what the main functions of genes and their products (proteins) are with regard to the organization of embryogenesis.

The source of the energy that drives the various cellular activities is not DNA, as has been believed for many years due to the authority of Charles Darwin; rather, it is the change in electrical charge [39].

Until the existing positions in the genetic code are re-examined, we will not be able to answer the question of where the coding functions of chromosomes originated and how they evolved.

The flaw in the principles of genetic coding lies in the excessive rigidity of the triplet-code model – which is viewed merely as a purely physicochemical system governing the operation of the ribosomal machinery – as well as in the unjustified attribution of all genetic attributes to this triplet-code framework. This misconception is based solely on the principle of complementarity in the interaction between the bases of messenger RNA (mRNA): adenine pairs with uracil, and guanine pairs with cytosine during the process of translating ribosomal information from mRNA. The protein code is perceived as some sort of universal, overarching vector that underlies all aspects of genetics and embryology. The genetic table became a kind of regularity. However, the genetic code has proven to be far from consistent; in fact, there are clear contradictions between the triplet-code model and the actual primary structure of proteins. Analysis of the genetic code reveals that, despite being a triplet-based system, the primary semantic burden lies with the duplet (base pair) at the beginning of each codon. There is reason to believe that this coding is in fact duplet-based. In 1979, it was discovered that mitochondria use a different genetic code; to date, 18 distinct protein codes have been identified for mitochondria and various organisms. Consequently, the genetic code is not universal.

The notion that a gene is a continuous segment of DNA corresponding to a single protein has been disproven. It has been found, however, that a single segment of DNA can contain information for two or even three proteins (as in the case of [108]). This indicates that the genetic code is overlapping; in other words, multiple proteins are encoded within the same gene.

It is noted that cells have sufficient reasons to use synonymous codons with varying frequencies. DNA mutations that replace codons with their synonyms do not result in changes to the amino acid sequence of the corresponding protein; such mutations are referred to as “silent” mutations. If a sense codon mutates into another sense codon that corresponds to a different amino acid, the resulting codon is referred to as a “reinterpreted” codon, or a codon with altered meaning (missense codon). Studies by several scientists have shown that, for example, the replacement of glutamic acid with valine can lead to a severe disease: sickle cell anemia.

The book devotes considerable attention to the selection of a mathematical model for the genome, in particular to a Markov chain model, using Bayesian recognition procedures. Based on this, an example of recognizing the secondary structure of proteins is provided.

It is noted that one of the main difficulties in analyzing nucleotide sequences is that the frequencies of adjacent nucleotides are not independent of each other. It has been found that the frequencies of pairs of adjacent bases usually differ from the product of the frequencies of those individual bases. These frequency patterns are related to the functional characteristics of the sequences. Regardless of the underlying reasons for the correlations between adjacent bases, studying these correlations is of great interest.

Attention has been paid to the presence of three reading frames within a single DNA strand. One of these reading frames is considered correct—the one in which the termination codons are located at the end of the strand. The other two reading frames are considered alternative when such termination codons occur within the strand. In the shifted states, completely different sequences of codons are obtained; each of these sequences will contain the termination codon “ter” multiple times. As is known, the codons ATG (AUG) and GTG (GUG) serve as initiation codons, while the three codons TAA (UAA), TAG (UAG), and TGA (UGA) do not correspond to any amino acid and indicate the end of the strand. However, it is unclear why there are three different termination codons. Although analyses of nucleotide sequences have shown that some genes end with two consecutive termination codons, there are also genes that use only one termination codon. The use of a second termination codon may serve as a kind of backup mechanism in the rare cases where the first codon becomes non-functional for some reason. However, as J. Watson pointed out, it is unclear why this mechanism is only employed in certain situations. Examples of sequences containing one or two termination codons are shown in Figure 8.3.

When forming the terminal codons in alternative reading frames, for example the codon TAA, the following amino acids are involved: Lysine (Lys), Asparagine (Asn) when the reading frame is shifted by 1 nucleotide to the left, and Isoleucine (Ile), Valine (Val), Leucine (Leu) when the reading frame is shifted by 1 nucleotide to the right. The stop codons TAG are formed with the participation of Arginine (Arg), Serine (Ser) when the reading frame is shifted by 1 nucleotide to the left, as well as Isoleucine (Ile), Valine (Val), Leucine (Leu). Finally, the stop codon TGA is formed with the involvement of Glutamate (Glu), Aspartate (Asp) when the reading frame is shifted by 1 nucleotide to the left, and Methionine (Met), Valine (Val), Leucine (Leu) when the reading frame is shifted by 1 nucleotide to the right.

mRNA encoding protein A of MS2

5' C G G C U C U C U A G A U A G A G C C C U C A A

mRNA encoding the coat protein of MS2 virus

5' U C C G G C A U C U A C U A A U A G A C G C C G

underlined = terminating (stop) codons

Figure 8.3: mRNA nucleotide sequences containing one or two stop codons

In addition, it can be observed that the amino acids Valine (Val) and Leucine (Leu), when the reading frame is shifted by one nucleotide, are involved in the formation of all three stop codons in one of the alternative reading frames.

The amino acids methionine and tryptophan are each coded by only one codon: ATG and TGG, respectively. The remaining amino acids that make up proteins are encoded by at least two codons, often four, and in some cases even six codons. Three amino acids (serine, leucine, and arginine) are coded by six codons each. These three sets of codon assignments are considered "irregular" compared to the seventeen other "regular" codons, for which the first and second positions in the codon sequences are identical. This characteristic makes the universal genetic code distinct from a uniform code, understood as a code with degeneracy only at the third base of the codon. Serine, leucine, and arginine are particularly complex in terms of their codon assignments; for example, arginine is coded by six different codons: AGA, AGG, CGA, CGG, CGT, and CGC. These synonymous codons are used by various organisms with varying frequencies. In humans, the codon AGG occurs in 22% of structural genes; in E. coli, it occurs in only 3%. The same is true for stop codons. For instance, the frequencies of the codons TAA, TAG, and TGA in human genes are 0.22, 0.17, and 0.61, respectively (meaning that TGA is used approximately 60% more frequently than TAA and TAG). In E. coli, these frequencies are 0.62, 0.09, and 0.30, respectively. Despite these differences, the genetic code is essentially the same across all organisms, with very few exceptions.

In the process of structural analysis, attention was paid to the search for direct repetitions within sequences of amino acid residues. The possibilities of obtaining information regarding the homology or partial homology of various sequences were also investigated. For example, the protein hormone adrenocorticotropin has a primary structure consisting of 39 amino acid residues. The first 24 residues are identical in all studied animals; however, the sequence of amino acid residues from positions 25 to 39 varies among different species [121].

When considering the Bayesian recognition procedure, it is noted that the inductive approach has advantages over the deductive one. The main advantage of the inductive approach is that the results are based on empirical information contained in all classes of the training dataset. Furthermore, unlike deductive systems, the processing of empirical data in inductive procedures is done in parallel for all classes. Finally, solving problems using inductive methods is easily feasible on modern computers.

Appendix 3 contains a program for one of the examples of Bayesian recognition using formula-based methods.

When examining logical methods for recognition, the role of features (attributes) is particularly noteworthy. If objects are described in terms of binary features and are subsequently represented as vectors in a Boolean space, the simplest way to construct a model is to identify patterns. The process of identifying these patterns is an inductive one: if a hypothesis is confirmed, it represents the desired pattern; if not, a new hypothesis must be formulated and tested again.

The number of possible object structures is quite large, and it needs to be decided how to represent the structure of such an object. A graph is a very convenient tool for describing the structure.

Table 5.1, presented in Chapter 5, allows for the quick identification of amino acids corresponding to a particular triplet. A group of Mexican researchers, led by Miguel Jiménez-Montano, proposed a method for visually representing triplets and their relationships in space. The publication of this method appeared in the journal "BioSystems" in 1996. Each base was characterized by the researchers using two independent properties.

1. By chemical class: R-purines (A, G), Y-pyrimidines (T, C).
2. The strength of the hydrogen bond: The letter "W" represents the weak hydrogen bond in the Watson-Crick base pair A=T, while the letter "S" represents the strong hydrogen bond in the G≡C base pair.

After examining these properties, the authors concluded that they could be represented by a binary number that can assume the values 0 or 1: the first digit of the binary number would indicate the chemical type (R or Y), while the second digit would indicate the strength of the hydrogen bond (W or S). The authors assigned the following binary values to the nitrogenous bases: A = 00; T = 10; G = 01; C = 11.

For the triplet code, a spatial structure was proposed that is isomorphic to the Boolean hypercube B_6 (see Figure 7.4, Section 7.4). In this hypercube, lines connect variables that differ from each other by exactly one bit of information.

To study the relationships and dependencies within the hypercube B_6 in more detail, special approaches are required. The set of amino acids can be considered as representing all the vertices of a unit six-dimensional cube.

The information required for the synthesis of proteins, which is encoded in genes, was first proposed by G. Gamow in 1954. He proposed the first model of the genetic code. Directly deciphering this code proved to be a challenging task; therefore, geneticists and mathematicians approached the problem indirectly. The question at hand was: how many nucleotides are needed to encode one amino acid? If only two nucleotides were sufficient for encoding an amino acid, there would be 16 possible combinations ($4 \times 4 = 16$). However, there are 20 amino acids in proteins, so the number of nucleotides required must be more than two. Three nucleotides per amino acid would result in $4 \times 4 \times 4 = 64$ possible combinations.

It has been established that the code falls into the category of "degenerate" codes; in other words, a particular amino acid can be encoded by one of several possible nucleotide triplets. But how can we determine where one triplet ends and another begins? For example, if there is a sequence of nucleotides such as ...TACTACTAC..., where T stands for thymine, A for adenine, and C for cytosine, should it be read as ...TAC TAC TAC..., or ...T ACT ACT ACT..., or perhaps even ...TA CTA CTA CTA...? We ultimately decided that the reading of this "genetic text" should start from a certain point, and then this text should be divided into three-letter "words". If the genetic "text" consists of such three-letter "words" that are to be read from one end only, then adding a single nucleotide at the beginning of the text will cause all subsequent triplets to change. Similarly, if two or more nucleotides are added either individually or in combination, the gene will not be able to function properly. A gene can only function when it contains exactly three types of mutations (additions or deletions of nucleotides), or in combinations of addition and deletion (one, two, etc.). In these cases, although the part of the gene between the added bases has undergone changes, the order of the bases in the remaining parts of the gene remains the same. In other words, the number of bases corresponding to one amino acid must be either equal to or a multiple of three.

Due to the rigid, planar conformation of peptide bonds, the conformation of proteins can be described using two dihedral angles located on the right and left sides of the alpha-carbon atoms (referred to as φ and ψ). Some researchers use this parameter, calling the dihedral angle a "torsional angle," and they predict the values of these torsional angles in protein amino acid sequences based on Bayesian recognition procedures using Markov chains. Experience has shown that Bayesian procedures using Markov chains of various orders can quite successfully predict the secondary structure of proteins.

Genomics is one of the main fields of modern molecular genetics. In 1978, F. Sanger and his colleagues sequenced the genome of the bacteriophage φX174, which contains a circular, single-stranded DNA molecule with a length of 5386 base pairs (bp); for this achievement, he was awarded the Nobel Prize in 1980. This milestone marked the beginning of a new era in molecular genetics: the field of genomics. In the late 1980s, the first international programs to sequence the complete genomes of bacteria, fungi, plants, insects, mammals, and humans were launched. Today, this research direction has evolved into bioinformatics—the science that deals with the technical methods for studying and modeling macromolecules and their systems.

Subsequently, in the phage φX174, a novel case was observed in which there were four overlapping genes. The discovery of these overlapping genes significantly expanded our understanding of the methods used in nature to encode and store genetic information. Further research revealed that the genomic organization characteristics found in phage φX174 are not unique at all. It is now believed that overlapping genes represent a somewhat unusual, but still quite common, form of genome organization in the organism [20].

The fundamental mysteries of living cells remain to be uncovered. We still do not know what causes a cell to divide, or what (or who) initially imparted life to it. Moreover, living systems are organized in such a way that many of their properties simply become inexplicable when tried to be explained by physics, chemistry, and other purely physical mechanisms. For instance, the basic functions of proteins and nucleic acids are not carried out through "pure chemistry" alone, but within the cellular context; they are realized in structures specific to life. Only by effectively integrating various scientific approaches can we gain new biological insights that will expand our understanding of the laws governing living matter. [23]

In further research, a major breakthrough should be the creation of a functional, complete mathematical model of a living cell. To develop such a model, the most advanced mathematical methods are required. We hope that in 35–40 years, scientists will have a precise mathematical theory that describes the interactions and dynamics of the highest levels of complexity in biological systems. For now, the key issue remains the origin of the standard genetic code.

APPENDIX 1

List of pairs that are formed by the genetic code:

```
        1 •
-1      Lys  +1
↓       AAA  ↓
   Lys AAA
   Glu GAA
  Stop TAA
   Gln CAA
          AAA Lys
          AAG Lys
          AAT Asn
          AAC Asn
```

```
        2 •
        Lys
        AAG
   Lys AAA
   Glu GAA
  Stop TAA
   Gln CAA
          AGA Arg
          AGG Arg
          AGT Ser
          AGC Ser
```

```
        3 •
        Asn
        AAT
   Lys AAA
   Glu GAA
  Stop TAA
   Gln CAA
          ATA Ile
          ATG Met
          ATT Ile
          ATC Ile
```

```
        4 •
        Asn
        AAC
   Lys AAA
   Glu GAA
  Stop TAA
   Gln CAA
          ACA Thr
          ACG Thr
          ACT Thr
          ACC Thr
```

```
        5 •
        Arg
        AGA
   Lys AAG
   Glu GAG
  Stop TAG
   Gln CAG
          GAA Glu
          GAG Glu
          GAT Asp
          GAC Asp
```

```
        6 •
        Arg
        AGG
   Lys AAG
   Glu GAG
  Stop TAG
   Gln CAG
          GGA Gly
          GGG Gly
          GGT Gly
          GGC Gly
```

```
        7 •
        Ser
        AGT
   Lys AAG
   Glu GAG
  Stop TAG
   Gln CAG
          GTA Val
          GTG Val
          GTT Val
          GTC Val
```

```
        8 •
        Ser
        AGC
   Lys AAG
   Glu GAG
  Stop TAG
   Gln CAG
          GCA Ala
          GCG Ala
          GCT Ala
          GCC Ala
```

```
        9 •
        Ile
        ATA
   Asn AAT
   Asp GAT
   Tyr TAT
   His CAT
          TAA Stop
          TAG Stop
          TAT Tyr
          TAC Tyr
```

```
        10 •
        Met
        ATG
   Asn AAT
   Asp GAT
   Tyr TAT
   His CAT
          TGA Stop
          TGG Trp
          TGT Tyr
          TGC Tyr
```

```
        11
        Ile
        ATT
   Asn AAT
   Asp GAT
   Tyr TAT
   His CAT
          TTA Leu
          TTG Leu
          TTT Phe
          TTC Phe
```

```
        12
        Ile
        ATC
   Asn AAT
   Asp GAT
   Tyr TAT
   His CAT
          TCA Ser
          TCG Ser
          TCT Ser
          TCC Ser
```

13
Thr
ACA
Asn AAC
Asp GAC
Tyr TAC
His CAC
CAA Gln
CAG Gln
CAT His
CAC His

14
Thr
ACG
Asn AAC
Asp GAC
Tyr TAC
His CAC
CGA Arg
CGG Arg
CGT Arg
CGC Arg

15
Thr
ACT
Asn AAC
Asp GAC
Tyr TAC
His CAC
CTA Leu
CTG Leu
CTT Leu
CTC Leu

16
Thr
ACC
Asn AAC
Asp GAC
Tyr TAC
His CAC
CCA Pro
CCG Pro
CCT Pro
CCC Pro

17 •
Glu
GAA
Arg AGA
Gly GGA
Stop TGA
Arg CGA
AAA Lys
AAG Lys
AAT Asn
AAC Asn

18 •
Glu
GAG
Arg AGA
Gly GGA
Stop TGA
Arg CGA
AGA Arg
AGG Arg
AGT Ser
AGC Ser

19 •
Asp
GAT
Arg AGA
Gly GGA
Stop TGA
Arg CGA
ATA Ile
ATG Met
ATT Ile
ATC Ile

20 •
Asp
GAC
Arg AGA
Gly GGA
Stop TGA
Arg CTA
ACA Thr
ACG Thr
ACT Thr
ACC Thr

21
Gly
GGA
Arg AGG
Gly GGG
Trp TGG
Arg CGG
GAA Glu
GAG Glu
GAT Asp
GAC Asp

22
Gly
GGG
Arg AGG
Gly GGG
Trp TGG
Arg CGG
GGA Gly
GGG Gly
GGT Gly
GGC Gly

23
Gly
GGT
Arg AGG
Gly GGG
Trp TGG
Arg CGG
GTA Val
GTG Val
GTT Val
GTC Val

24
Gly
GGC
Arg AGG
Gly GGG
Trp TGG
Arg CGG
GCA Ala
GCG Ala
GCT Ala
GCC Ala

25 •
Val
GTA
Ser AGT
Gly GGT
Cys TGT
Arg CGT
TAA Stop
TAG Stop
TAT Tyr
TAC Tyr

26 •
Val
GTG
Ser AGT
Gly GGT
Cys TGT
Arg CGT
TGA Stop
TGG Trp
TGT Cys
TGC Cys

27
Val
GTT
Ser AGT
Gly GGT
Cys TGT
Arg CGT
TTA Leu
TTG Leu
TTT Phe
TTC Phe

28
Val
GTC
Ser AGT
Gly GGT
Cys TGT
Arg CGT
TCA Ser
TCG Ser
TCT Ser
TCC Ser

29
Ala
GCA
Ser AGC
Gly GGC
Cys TGC
Arg CGC
CAA Gln
CAG Gln
CAT His
CAC His

30
Ala
GCG
Ser AGC
Gly GGC
Cys TGC
Arg CGC
CGA Arg
CGG Arg
CGT Arg
CGC Arg

31
Ala
GCT
Ser AGC
Gly GGC
Cys TGC
Arg CGC
CTA Leu
CTG Leu
CTT Leu
CTC Leu

32
Ala
GCC
Ser AGC
Gly GGC
Cys TGC
Arg CGC
CCA Pro
CCG Pro
CCT Pro
CCC Pro

33
Stop
TAA
Ile ATA
Val GTA
Leu TTA
Leu CTA
AAA Lys
AAG Lys
AAT Asn
AAC Asn

34
Stop
TAG
Ile ATA
Val GTA
Leu TTA
Leu CTA
AGA Arg
AGG Arg
AGT Ser
AGC Ser

35
Tyr
TAT
Ile ATA
Val GTA
Leu TTA
Leu CTA
ATA Ile
ATG Met
ATT Ile
ATC Ile

36
Tyr
TAC
Ile ATA
Val GTA
Leu TTA
Leu CTA
ACA Thr
ACG Thr
ACT Thr
ACC Thr

37
Stop
TGA
Met ATG
Val GTG
Leu TTG
Leu CTG
GAA Glu
GAG Glu
GAT Asp
GAC Asp

38
Trp
TGG
Met ATG
Val GTG
Leu TTG
Leu CTG
GGA Gly
GGG Gly
GGT Gly
GGC Gly

39
Cys
TGT
Met ATG
Val GTG
Leu TTG
Leu CTG
GTA Val
GTG Val
GTT Val
GTC Val

40
Cys
TGC
Met ATG
Val GTG
Leu TTG
Leu CTG
GCA Ala
GCG Ala
GCT Ala
GCC Ala

41 •
Leu
TTA
Ile ATT
Val GTT
Phe TTT
Leu CTT
TAA Stop
TAG Stop
TAT Tyr
TAC Tyr

42 •
Leu
TTG
Ile ATT
Val GTT
Phe TTT
Leu CTT
TGA Stop
TGG Trp
TGT Cys
TGC Cys

43
Phe
TTT
Ile ATT
Val GTT
Phe TTT
Leu CTT
TTA Leu
TTG Leu
TTT Phe
TTC Phe

44
Phe
TTC
Ile ATT
Val GTT
Phe TTT
Leu CTT
TCA Ser
TCG Ser
TCT Ser
TCC Ser

45
Ser
TCA
Ile ATC
Val GTC
Phe TTC
Leu CTC
CAA Gln
CAG Gln
CAT His
CAC His

46
Ser
TCG
Ile ATC
Val GTC
Phe TTC
Leu CTC
CGA Arg
CGG Arg
CGT Arg
CGC Arg

47
Ser
TCT
Ile ATC
Val GTC
Phe TTC
Leu CTC
CTA Leu
CTG Leu
CTT Leu
CTC Leu

48
Ser
TCC
Ile ATC
Val GTC
Phe TTC
Leu CTC
CCA Pro
CCG Pro
CCT Pro
CCC Pro

49
Gln
CAA
Thr ACA
Ala GCA
Ser TCA
Pro CCA
AAA Lys
AAG Lys
AAT Asn
AAC Asn

50
Gln
CAG
Thr ACA
Ala GCA
Ser TCA
Pro CCA
AGA Arg
AGG Arg
AGT Ser
AGC Ser

51
His
CAT
Thr ACA
Ala GCA
Ser TCA
Pro CCA
ATA Ile
ATG Met
ATT Ile
ATC Ile

52
His
CAC
Thr ACA
Ala GCA
Ser TCA
Pro CCA
ACA Thr
ACG Thr
ACT Thr
ACC Thr

53
Arg
CGA
Thr ACG
Ala GCG
Ser TCG
Pro CCG
GAA Glu
GAG Glu
GAT Asp
GAC Asp

54
Arg
CGG
Thr ACG
Ala GCG
Ser TCG
Pro CCG
GGA Gly
GGG Gly
GGT Gly
GGC Gly

55
Arg
CGT
Thr ACG
Ala GCG
Ser TCG
Pro CCG
GTA Val
GTG Val
GTT Val
GTC Val

56
Arg
CGC
Thr ACG
Ala GCG
Ser TCG
Pro CCG
GCA Ala
GCG Ala
GCT Ala
GCC Ala

57 •
Leu
CTA
Thr ACT
Ala GCT
Ser TCT
Pro CCT
TAA Stop
TAG Stop
TAT Tyr
TAC Tyr

58 •
Leu
CTG
Thr ACT
Ala GCT
Ser TCT
Pro CCT
TGA Stop
TGG Trp
TGT Cys
TGC Cys

59
Leu
CTT
Thr ACT
Ala GCT
Ser TCT
Pro CCT
TTA Leu
TTG Leu
TTT Phe
TTC Phe

60
Leu
CTC
Thr ACT
Ala GCT
Ser TCT
Pro CCT
TCA Ser
TCG Ser
TCT Ser
TCC Ser

```
    61
    Pro
    CCA
Thr ACC
Ala GCC
Ser TCC
Pro CCC
      CAA Gln
      CAG Gln
      CAT His
      CAC His
```

```
    62
    Pro
    CCG
Thr ACC
Ala GCC
Ser TCC
Pro CCC
      CGA Arg
      CGG Arg
      CGT Arg
      CGC Arg
```

```
    63
    Pro
    CCT
Thr ACC
Ala GCC
Ser TCC
Pro CCC
      CTA Leu
      CTG Leu
      CTT Leu
      CTC Leu
```

```
    64

    Pro
    CCC
Thr ACC
Ala GCC
Ser TCC
Pro CCC
      CCA Pro
      CCG Pro
      CCT Pro
      CCC Pro
```

APPENDIX 2

Amino acids as fragments of a polypeptide chain

Aliphatic					Sulfur-containing	
glycine (Gly, G)	alanine (Ala, A)	valine * (Val, V)	leucine * (Leu, L)	isoleucine * (Ile, I)	cysteine (Cys, C)	methionine * (Met, M)
H *polarity*	CH_3	H_3C-CH CH_3	CH_2 H_3C-CH CH_3	H_3C-C-H CH_2 CH_3	CH_2 SH pK_R 8.3	CH_2 CH_2 S CH_3
−2.4	−1.9	−2.0	−2.3	−2.2	−1.2	−1.5

Aromatic			Imino acids	Neutral	
phenylalanine * (Phe, F)	tyrosine (Tyr, Y)	tryptophan * (Trp, W)	proline (Pro, P)	serine (Ser, S)	threonine * (Thr, T)
CH_2 	CH_2  OH 10	CH_2 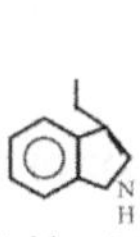	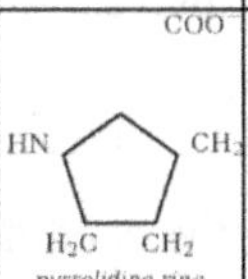	CH_2 OH	H_3C-C-H OH
+0.8	+6.1	+5.9	+6.0	+5.1	+4.9
* essential amino acids					chiral centre

Neutral		Acidic		Basic		
asparagine (Asn, N)	glutamine (Gln, Q)	aspartic acid (Asp, D)	glutamic acid (Glu, E)	histidine (His, H)	lysine * (Lys, K)	arginine * (Arg, R)
CH_2 $CONH_2$	CH_2 CH_2 $CONH_2$	CH_2 COO^- 4.0	CH_2 CH_2 COO^- 4.3	CH_2 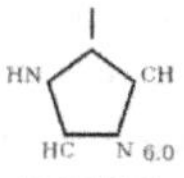	CH_2 CH_2 CH_2 CH_2 N^+H_3 10.8	CH_2 CH_2 CH_2 NH 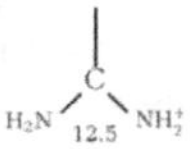
+9.7	+9.4	+11.0	+10.2	+10.3	+15.0	+20.0

APPENDIX 3

A Pascal-language program for the recognition example

Example of a Bayesian recognition procedure

Let's consider the application of Bayes' formula for recognizing objects that belong to three classes. The original set U can be represented as a set of binary words of a specified length: n = 6 (e.g., x_1, x_2, x_3, x_4, x_5, x_6). Therefore, the recognition will be performed based on six general features: x_1, x_2, x_3, x_4, x_5, x_6. The number of such binary words (objects) will be equal to 2^6, which equals 64. For each of the 64 objects, there are multiple values corresponding to three different classes of characteristics. The recognized objects must assume one of three values (classes): A_1, A_2, A_3. In order to distinguish between the description data, it is necessary to assign a probability distribution to each of the object descriptions. This should be done for all objects in the dataset. Bayes' formula allows us to generate such a distribution and determine the membership of objects to their respective classes. As we have already mentioned, the Bayesian method allows us to successfully take into account prior information in many cases. The prior probabilities of $p(A_i)$ are known or can be determined based on previous experiments, for various reasons.

The Bayes formula is written as follows:

$$P(A_s|x_1, x_2, \dots, x_n) = \frac{P(A_s)\cdot P(x_1|A_s)\cdot P(x_2|A_s)\cdot \dots \cdot P(x_n|A_s)}{P(x_1,\dots,x_j,\dots,x_n)} \quad \text{(A.3.1)}$$

The probability entering the denominator of this formula

$$P(x_1, \dots, x_j, \dots, x_n) = \sum_i^s P(A_s) \cdot \prod_j^6 P(x_j|A_s) \quad \text{(A.3.2)}$$

This is referred to as the "full probability."

Where:

- s represents the number of classes;
- n represents the number of features (in our case, n = 6);
- As is one of the classes (along with A_1, A_2, A_3);
- $P(x_j|A_s)$ represents the conditional probabilities of a feature being present in class A_s;
- $P(A_1)$, $P(A_2)$, $P(A_3)$ represent the prior probabilities, determined based on previous experiments. Let's assume they are distributed as follows: $p(A_1) = 0.25$; $p(A_2) = 0.4$; $p(A_3) = 0.35$.

We assume that the features are independent of each other. It's important to note that not only the presence of a feature provides information about the class, but so does its absence.

From the theory of probability, it is known that the probability of the opposite event is…

The expression x_iA_j can be rewritten as $p(x_i|A_j) = 1 - p(\bar{x}i|A_j)$. (A.3.3)

So, the Bayesian recognition procedure works based on samples, which are used to determine the classification of objects into specific categories. Let the researcher determine the classification of the objects into specific categories based on the working samples. These are sets of data (in our example, working samples from 64 sets in total), on which the functions for each class take values equal to 1. The sets in which the functions accept values of 0 for the corresponding classes will be referred to as zeroed-out or disabled sets. In order to find the function of a specific class f, it is sufficient to list all the singleton sets and combine them using the disjunction operator. In order to achieve a more compact recording format, efforts are made to minimize the use of Boolean functions. There are several ways to minimize things to the greatest extent possible. One of these methods is the Karnaugh map technique, which is considered the most convenient approach for quickly solving problems involving the minimization of Boolean functions with a sufficiently large number of arguments (6–12). As a result, the minimized function is obtained in the form based on the logical operators AND, OR, and NOT. Here, we will not go into detail about the Karnaugh map method; instead, we will only mention that we will use the apostrophe "'" to represent negation, along with "[141]". It turns out that the objects belonging to class A_1 satisfy one of the following conditions: - $x_1 = 0$, $x_2 = 0$, $x_6 = 1$; - or $x_1 = 0$, $x_2 = 1$; - or $x_1 = 1$, $x_2 = 1$, $x_6 = 1$. The class A_2 includes objects that satisfy one of the following conditions: - $x_1 = 1$ and $x_5 = 0$, $x_6 = 0$; - or $x_1 = 1$ and $x_2 = 0$, $x_5 = 0$; - or $x_1 = 0$ and $x_2 = 0$, $x_6 = 0$. The class A_3 includes objects that satisfy the following conditions: $x_1 = 1$ and $x_5 = 1$. As a result of the minimization process, we obtain the assignment of functions to their respective classes: To class A_1: x_1', x_2', x_6; x_1', x_2; x_1, x_2, x_6. To the class A_2: x_1, x_5', x_6'; x_2', x_5'; x_1', x_2', x_6'. To class A_3: x_1, x_5.

All of these model characteristics must be taken into account when preparing the system for training. The recognition system, during its development phase, works together with a "teacher." The teacher's role is to present the system with selected objects and indicate which classes those objects belong to. Afterwards, the teacher tests the recognition system by providing it with control objects. If the system correctly identifies the control objects, it means that the system has learned how to perform recognition tasks. If necessary, the teacher increases the size of the training dataset until the average number of errors falls below a desired level. In this way, the system gains the ability to detect very subtle differences between the recognized objects. By "datasets," we mean explicitly defined, non-overlapping subsets of representatives from the classes A_1, A_2, and A_3 (denoted as E_i). Our goal is to find algorithms that can determine for any object in the space U to which class it belongs, with the condition that all elements in the dataset must remain within their respective classes.

The initial set U, consisting of the 64 binary six-digit words, is represented as the table below (Fig. 4), in which all three classes of objects are marked out. Class A_1 comprises the objects placed in the white cells; class A_2, the objects in the dark-gray cells; and class A_3, the objects in the light-gray cells.

	1	2	3	4	5	6	7	8
1	000000	000100	000110	000010	000011	000111	000101	000001
2	001000	001100	001110	001010	001011	001111	001101	001001
3	011000	011100	011110	011010	011011	011111	011101	011001
4	010000	010100	010110	010010	010011	010111	010101	010001
5	110000	110100	110110	110010	110011	110111	110101	110001
6	111000	111100	111110	111010	111011	111111	111101	111001
7	101000	101100	101110	101010	101011	101111	101101	101001
8	100000	100100	100110	100010	100011	100111	100101	100001

Figure 4. Classification of all 64 six-digit binary words into the three classes A_1, A_2, A_3, laid out with the symmetric Gray code.

Legend: A_1 — white; A_2 — dark gray; A_3 — light gray.

Given any way of dividing a set into classes (i.e., any partition of the set into subsets), we can construct an algorithm that solves this problem. In the training mode, conditional probabilities are determined; the set in question is also used as part of the algorithm's input data.

The conditional probabilities $p(x_i|A_j)$ and $p(x_i|A_j)$ serve as prototypes for the class A_j. The training of the system involves continuously refining the probabilities used in the calculations. The input data (samples) for calculating the conditional probabilities $p(\bar{x}j|A_i)$ are represented in Tables A.3.1, A.3.2, and A.3.3.

Table A.3.1

This applies to the class A_1.

Object ID:	x_1	x_2	x_3	x_4	x_5	x_6
1	0	0	0	1	0	1
2	0	1	0	0	0	0
3	0	0	1	0	1	1
4	1	0	1	1	0	1
5	1	1	1	0	0	1

Table A.3.2

This applies to the class A_2.

Object ID:	x_1	x_2	x_3	x_4	x_5	x_6
6	1	0	0	0	0	0
7	1	0	1	0	0	0
8	1	0	1	0	0	1
9	1	0	0	1	0	1
10	0	1	1	1	1	0

Table A.3.3

This applies to the class A_3.

Object ID:	x_1	x_2	x_3	x_4	x_5	x_6
11	1	0	1	0	1	1
12	1	0	0	0	1	0
13	1	0	1	1	0	0
14	1	0	0	1	1	0
15	1	1	0	0	1	1

As can be seen from the tables, the conditional probabilities for objects with the corresponding class values: A_1, A_2, A_3, are distributed as follows: _

$P(x_1 \mid A_1) = 0.4$; $P(\bar{x}1 \mid A_1) = 0.6$;

$P(x_2 \mid A_1) = 0.4$; $P(\bar{x}2 \mid A_1) = 0.6$;

$P(x_3 \mid A_1) = 0.6$; $P(\bar{x}3 \mid A_1) = 0.4$;

$P(x_4 \mid A_1) = 0.4$; $P(\bar{x}4 \mid A_1) = 0.6$;

$P(x_5 \mid A_1) = 0.2$; $P(\bar{x}5 \mid A_1) = 0.8$;

$P(x_6 \mid A_1) = 0.8$; $P(\bar{x}6 \mid A_1) = 0.2$;

$P(x_1 \mid A_2) = 0.8$; $P(\bar{x}1 \mid A_2) = 0.2$;

$P(x_2 \mid A_2) = 0.2$; $P(\bar{x}2 \mid A_2) = 0.8$;

$P(x_3 \mid A_2) = 0.6$; $P(\bar{x}3 \mid A_2) = 0.4$;

$P(x_4 \mid A_2) = 0.4$; $P(\bar{x}4 \mid A_2) = 0.6$;

$P(x_5 \mid A_2) = 0.2$; $P(\bar{x}5 \mid A_2) = 0.8$;

$P(x_6 \mid A_2) = 0.4$; $P(\bar{x}6 \mid A_2) = 0.6$;

$P(x_1 \mid A_3) = 1$; $P(\bar{x}1 \mid A_3) = 0$;

$P(x_2 \mid A_3) = 0.2$; $P(\bar{x}2 \mid A_3) = 0.8$;

$P(x_3 \mid A_3) = 0.4$; $P(\bar{x}3 \mid A_3) = 0.6$;

$P(x_4 \mid A_3) = 0.4$; $P(\bar{x}4 \mid A_3) = 0.6$;

$P(x_5 \mid A_3) = 0.8$; $P(\bar{x}5 \mid A_3) = 0.2$;

$P(x_6 \mid A_3) = 0.4$; $P(\bar{x}6 \mid A_3) = 0.6$.

The schematic diagram of the recognition system, which operates using Bayes' formula, is shown in Figure A.3.1.

For example, suppose you need to classify an object that is characterized by a set of parameters: $x_1 = 1$, $x_2 = 1$, $x_3 = 0$, $x_4 = 1$,, $x_5 = 0$, $x_6 = 1$. Based on the conditional probabilities associated with these values (h_j), the system performs the following calculation: _ _

$P(A_1)\times P(x_1|A_1)\times P(x_2|A_1)\times P(x_3|A_1)\times P(x_4|A_1)\times P(x_5|A_1)\times P(x_6|A_1) =$
$0.25\times 0.4\times 0.4\times 0.4\times 0.4\times$

$0.8\times 0.8 = 0.0041;$

$P(A_2)\times P(x_1|A_2)\times P(x_2|A_2)\times P(x_3|A_2)\times P(x_4|A_2)\times P(x_5|A_2)\times P(x_6|A_2) =$
$0.4\times 0.8\times 0.2\times 0.4\times 0.4\times 0.8\times 0.4 = 0.0033;$

$P(A_3)P(x_1|A_3)P(x_2|A_3)P(x_3|A_3)(x_4|A_3)P(x_5|A_3)P(x_6|A_3) =$

$= 0.35\times 1\times 0.2\times 0.6\times 0.4\times 0.2\times 0.4 = 0.0013.$

Next, we have: $P(x_1, \dots, x_j, \dots, x_n) = \sum_i^s P(A_i) \cdot \prod_j^6 P(x_j|A_i) =$

$= 0.0041 + 0.0033 + 0.0013 = 0.0087.$

The final result is: $P(A1|x_1, x_2, x_3, x_4, x_5, x_6) = \frac{0.0041}{0.0087} = 0.47;$

$$P(A2|x_1, x_2, x_3, x_4, x_5, x_6) = \frac{0.0033}{0.0087} = 0.38;$$

$$P(A3|x_1, x_2, x_3, x_4, x_5, x_6) = \frac{0.0013}{0.0087} = 0.15.$$

By calculating the probabilities for the objects to be recognized in this manner, the system compares them and indicates the value among the three options for which the probability is the highest. In our example, the comparator points to the value A_1.

It should be noted that after training, the system acquired the ability to detect very subtle differences among the recognized objects. For example, the objects with numbers 10 and 13 were "incorrectly" assigned to the classes specified in the tables due to their slight distinctions from neighboring classes. The system was able to respond correctly and assign these objects to their proper classes: object number 10 to class A_1 and object number 13 to class A_2.

Let's send a few more objects to the system's input and determine the probabilities of their belonging to the respective classes:

For an object with the parameter set: 1, 1, 0, 0, 0, 1, the system outputs the following values: $p(A_1) = 0.4699$; $p(A_2) = 0.3759$; $p(A_3) = 0.1542$.

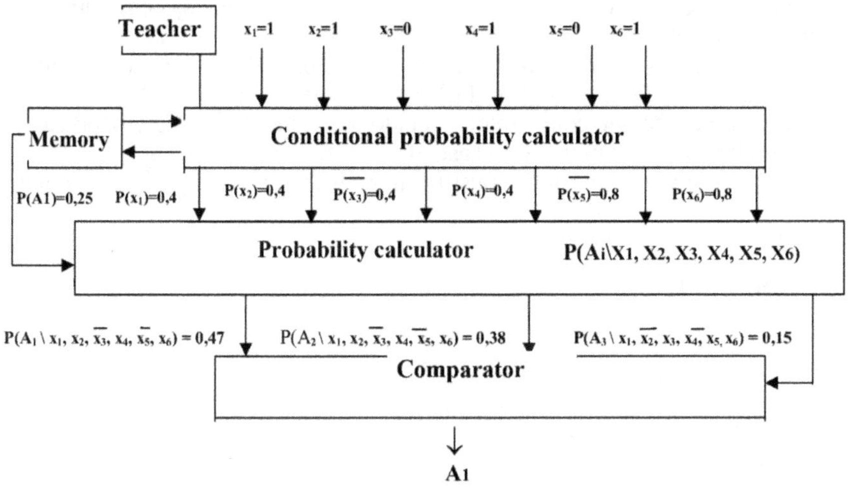

Fig. A.3.1. Block diagram of the recognition system operating via Bayes' formula

For an object with the parameter set: 1, 1, 1, 0, 0, 0, the system outputs the following values: $p(A_1) = 0.1498$; $p(A_2) = 0.7191$; $p(A_3) = 0.1311$.

For an object with the parameter set: 1, 0, 1, 0, 1, 0, the system outputs the following values: $P(A_1) = 0.0196$; $P(A_2) = 0.2503$; $P(A_3) = 0.7301$.

In the first case, the comparator refers to the value A_1; in the second case, it refers to the value A_2; in the third case, it refers to the value A_3.

Now, for this example, let's construct a program for a recognition system that operates using Bayes' formula.

The program is written in Pascal.

```
Program analys;
uses crt;
const
a1:array[1..5,1..6] of
byte=((0,0,0,1,0,1),(0,1,0,0,0,0),(0,0,1,0,1,1),
(1,1,1,1,0,1),(1,1,1,0,0,1));
 a2:array[1..5,1..6] of
byte=((1,0,0,0,0,0),(1,0,1,0,0,0),(1,0,1,0,0,1),
(1,0,0,1,0,1),(0,1,1,1,1,0));
 a3:array[1..5,1..6] of
byte=((1,0,1,0,1,1),(1,0,0,0,1,0),(1,0,1,1,0,0),
(1,0,0,1,1,0),(1,1,0,0,1,1)),
ver1:array[1..6] of real=(0.4,0.4,0.6,0.4,0.2,0.8);
ver11:array[1..6] of real=(0.6,0.6,0.4,0.6,0.8,0.2);
ver2:array[1..6]of real=(0.8,0.2,0.6,0.4,0.2,0.4);
ver22:array[1..6] of real=(0.2,0.8,0.4,0.6,0.8,0.6);
ver3:array [1..6] of real=(1.0,0.2,0.4,0.4,0.8,0.4);
ver33:fray [1..6] of real=(0.0,0.8,0.6,0.6,0.2,0.6);
pa1=0.25;
pa2=0.4;
pa3=0.35;
var
x:array[1..6] of byte;
i,j:byte;
max,b1,b2,b3:real;
begin
clrscr;
gotoxy(35,wherey);
writeln('Diagnostica');
writeln;
for i:=1 to 5 do
begin
for:=1 to 6 do
write(a1[i,j]:3,' ');
writeln;
end;
for i:=1 to 6 do
write(ver1[i]:1:1,' ');
 writeln;
 writeln;
```

```
for i:=1 to 6 do
write(ver11[i]:1:1,' ');
writeln;
gotoxy(1,3);
for i:=1 to5 do
begin
gotoxy(26,wherey);
for j:=1 to 6 do
write(a2[i,j]:3,' ');
writeln;
end;
gotoxy(26,wherey);
for i:=1 to 6 do
write(ver2[i]:1:1,' ');
writeln;
writeln;
gotoxy(26,wherey);
for i:=1 to 6 do
write(ver22[i]:1:1,' ');
writeln;
gotoxy(1,3);
for i:=1 to 5 do
begin
gotoxy(52,wherey);
for j:=1 to 6 do
     wrie(a3[i,j]:3,'  ');
writeln;
end;
qotoxy(52, wherey);
for i:= 1 to6 do
write(ver3[i]:1:1,' ');
writeln;
writeln;
qotoxy(52,wherey);
for i:=1 to 6 do
write(ver33[i]:1:1,' ');
writeln;
writeln('Nabor haracteristic');
for i:=1 to6 do
begin
write('x', i:1,'-');
readin(x[i]);
end;
b1:=pa 1;
for i:=1 to6 do
if x[i]=1 then
b1:=b1*ver1[i]
else
b1:=b1*ver11[i];
b2:=pa2;
```

```
for i:1 to 6 do
if x[i]= 1 then
b2:=b2*ver2[i]
else
b2:=b2*ver22[i];
b3:=pa3;
for i:= 1 to 6 do
if [i]=1 then
b3:=b3*ver3[i]
else
b3:=b3*ver33[i];
max:=b1/(b1+b2+b3);
if b2>max then
max:=b2/(b1+b2+b3);
if b3>max then
max:= b3/(b1+b2+b3);
writeln('Resultat');
write('A1=',b1/(b1+b2+b3):1:4,'A2=',b2/(b1+b2+b3):1:4,'A3=',
b3/(b1+b2+b3):1:4);
writeln;
writeln('maximum',max:1:4);
repeat until keypressed;
end.
```

TERMINOLOGY DICTIONARY

A

A posteriori — something that arises from experience; a concept in the theory of knowledge, as opposed to a priori.

A priori — a concept in logic and epistemology that refers to knowledge that precedes experience and is independent of it.

adenine — a purine base that is part of both DNA and RNA; it is complementary to thymine and uracil. It is one of the four "letters" that make up the genetic code.

adenosine triphosphate (ATP) — a nucleotide composed of adenosine and three phosphate groups. In all living organisms, it functions as a universal energy storage molecule.

alignment — a method used to obtain both an analytical and a graphical representation of the pattern underlying a given empirical sequence of statistical data. Two sequences are considered to be aligned if they are arranged in such a way as to maximize their similarity to each other.

alleles — different forms of the same gene, located in the same regions (loci) on homologous (paired) chromosomes; they determine various manifestations of the same trait. In a normal diploid cell, no more than two alleles can exist simultaneously.

amino acids — a class of organic compounds that contain carboxyl groups (COOH) and amino groups (NH_2); they possess the properties of both acids and bases. The nitrogen-containing compounds involved in the exchange between all organisms are the starting materials for the biosynthesis of hormones, vitamins, pigments, as well as purine and pyrimidine bases. Approximately 20 standard (proteinogenic) amino acids act as the monomer units from which all proteins are constructed; the order in which these amino acids are arranged within the proteins is determined by the genetic code.

amino group — NH2 a monovalent atomic group; it corresponds to the functional group of primary amines (RNH_2). This group is also part of the structure of amino acids.

anticodon — A segment of transfer RNA (tRNA) that consists of three nucleotides. It specifically binds to a codon (a sequence of three nucleotides on messenger RNA (mRNA)) to ensure the correct placement of each amino acid in the polypeptide chain during protein synthesis.

antigens — complex organic substances that, upon entering the bodies of animals and humans, can trigger an immune response, resulting in the production of antibodies.

assimilation — (from Latin assimilatio - meaning "to use or incorporate") is the process by which an organism takes in and utilizes substances that enter from the external environment.

atomic mass (atomic weight) — the mass of an atom of a chemical element, expressed in atomic mass units.

B

bacterium — a single-celled microorganism that belongs to the "prokaryotic" forms of life, which are organisms without a nucleus.

Bayesian estimation — the process of estimating an unknown parameter based on observed data, using the Bayesian approach.

bayes' criterion — a rule for selecting a decision strategy that minimizes the average risk.

bioinformatics — the science that studies biological information using mathematical, statistical, and computational methods.

Boolean function in the algebra of logic — a function whose arguments, as well as the function itself, take values from a two-element set (usually {0, 1}).

C

carboxyl group (COOH) — a monovalent functional group consisting of a carboxyl group and a hydroxyl group; it determines the acidic properties of carboxylic acids.

cell — a fundamental living system that constitutes the basis for the structure and functioning of all animals and plants.

chaperones — (from french "Chaperon" meaning "nanny") protein complexes that facilitate the process of folding newly synthesized polypeptides into their correct three-dimensional structures by restricting their interactions with other proteins.

chromosomes — structural elements of the cell nucleus that contain DNA, which carries the genetic information of an organism. Genes are arranged in a linear sequence within chromosomes. The process of self-doubling and the orderly distribution of chromosomes to daughter cells during cell division ensure the transmission of an organism's genetic traits from one generation to the next.

codon (a trimer of nucleotides) — the unit of genetic code; it consists of three consecutive nucleotides in a DNA or RNA molecule. The sequence of codons in a gene determines the sequence of amino acids in the polypeptide chain of the protein encoded by that gene. Of the 64 possible codons, 61 encode 20 different amino acids (some amino acids can be coded by multiple codon synonyms), while the remaining three serve as "stop signals" that terminate the polypeptide synthesis process.

comparator — (from latin comparator, meaning "to compare") a device used to measure a quantity and compare it with a standard. There are various types of comparators, such as optical and electrical comparators. These devices are used, for example, to measure the intensity of electromagnetic fields, etc. In astronomy (for comparing spectra and astronomical photographs), instruments like spectrocomparators and blink comparators are employed.

complementarity — the mutual correspondence in the chemical structure of two macromolecules that enables their interaction—such as the pairing of two strands of DNA, the binding of an enzyme to its substrate, or the binding of an antigen to an antibody. Complementary structures fit together precisely, like a key fitting into a lock.

conditional probability — In some cases, we need to consider the probabilities of events under the additional condition that another event B has occurred. Such probabilities are called conditional probabilities and are denoted by the symbol $P(A|B)$; this means the probability of event A occurring given that event B has taken place.

conformations — (from Latin conformatio - meaning "shape, arrangement") refer to the geometric configurations that molecules of organic compounds adopt when individual parts of the molecule can rotate around single bonds. Compounds that differ only in conformation are called rotational isomers (conformers); under normal conditions these readily interconvert, and therefore, in the case of simple compounds, they generally cannot be separated.

correlation (in mathematical statistics) — a probabilistic or statistical dependence between two variables. Unlike a functional dependence, correlation arises when the relationship between two features is complicated by the presence of various random factors.

cytosine — a pyrimidine base. It is found in all living organisms as part of nucleic acids (and is one of the four "letters" that make up the genetic code).

D

database — a specialized collection of data that is accumulated and stored as the primary information source for use in a particular field of science, practice, or education. It is a collection of data organized in a way that facilitates easy and rapid retrieval of the required information.

deduction — (from latin deductio meaning "to draw out") is a logical process in which general statements are used to derive specific conclusions from given premises. If the premises of a deduction are true, then the conclusion must also be true.

deletion — (from latin "deletio" meaning destruction) the loss of one of the internal segments of a chromosome.

deoxyribonucleic acid (DNA) — a high-molecular-weight natural compound found in the nuclei of cells of living organisms. DNA carries genetic information; its individual segments correspond to specific genes. Combinations of three adjacent nucleotides in the DNA chain (called codons) constitute the genetic code. Disturbances in the sequence of nucleotides in the DNA chain lead to hereditary changes in the organism—mutations.

deoxyribose — a simple sugar that contains one fewer hydroxyl group than ribose.

disjunction — (from latin disjunctio meaning "separation") the logical equivalent of the "or" conjunction; it is an operation that formalizes the main logical properties of this conjunction.

dissimilation — (from latin dissimilis meaning "different") the process of breaking down organic substances into simpler compounds, thereby releasing energy necessary for the organism's vital functions.

domain — a portion of an amino acid sequence in a protein that is associated with a specific function.

duplication (biology) — The same as "replication.".

duplication — (from the Latin "duplicatio" meaning "to double") is a type of chromosomal rearrangement (mutation) that involves the duplication of a certain portion of a chromosome.

E

embryogenesis — (from "embryo" and "genesis") the same process as embryonic development.

enzymes — (from the Latin "fermentum" – meaning yeast) are biological catalysts that are present in all living cells. They facilitate chemical reactions within an organism, thereby directing and regulating its metabolism. Chemically speaking, enzymes are proteins that exhibit optimal activity levels. Enzymatic activity depends on optimal environmental conditions (temperature, pH level) and is inhibited by various chemical agents belonging to the class of inhibitors.

epigenetics (or epigenomics) — studies the formation and hereditary transmission of a specific functional state of the genome, including the factors that regulate transcription (the process of translating genetic information into protein). Among these factors, methylation (the addition of a methyl group to cytosine in DNA), acetylation (the transfer of acetyl groups to lysine residues in histone molecules), and other similar modifications play particular roles.

Error-correcting codes — those that enhance reliability in the event of random errors. These codes can be divided into two categories: those that detect errors and those that actually fix them.

eukaryotes — (from the Greek "eu" – good, complete, and "karyon" – nucleus) are organisms (all animals, most plants) that, unlike prokaryotes, possess a well-defined cell nucleus enclosed within a nuclear membrane that separates it from the cytoplasm. The genetic material is contained within chromosomes.

exon — the expressed (i.e., actually translated into a protein) portion of a gene; it represents the coding region of the gene that is retained in the mature mRNA molecule.

expert evaluations — Quantitative and/or ordinal assessments of processes or phenomena that cannot be directly measured. These evaluations are based on the judgments of experts.

expressiveness — (genetic) the degree to which a trait is expressed, as determined by the gene. It can vary depending on the genotype in which the gene is present, as well as on the conditions of the external environment.

F

folding — the process of forming the three-dimensional (spatial) structure of proteins. This folding is carried out by specialized protein complexes.

functional — a mathematical concept that emerged in the calculus of variations to denote a variable quantity defined on a set of functions, meaning it depends on the choice of one or more functions. For example, the length of the arc connecting two fixed points is a functional, because the length of the arc depends on the function whose graph connects those points.

G

gametes — reproductive cells that contain a haploid (uniform) set of chromosomes and are involved in sexual reproduction. When two gametes fuse during the sexual process, a zygote is formed.

gene — the unit of hereditary material responsible for the formation of some elementary trait. In higher organisms (eukaryotes), it is part of chromosomes. The collection of all genes in an organism constitutes its genetic makeup — the genotype. Each gene is responsible for the synthesis of a specific protein (an enzyme, etc.).

gene expression — the active state of a gene, during which the genetic information contained within it is translated into reality: the synthesis of RNA and proteins. The expression of genes can be regulated at all stages of the process – during transcription, translation, as well as during post-translational modifications of the proteins.

gene mapping — the process of determining the location of a particular gene on a chromosome in relation to other genes.

gene pool — the collection of genes present in the individuals making up a given population.

genetic code — a unique system inherent in living organisms for "recording" hereditary information in nucleic acid molecules, in the form of a sequence of nucleotides. The unit of the genetic code is the codon (a triplet of nucleotides). The implementation of the genetic code within a cell occurs in two stages: 1. Transcription: This process takes place in the nucleus and involves the synthesis of messenger RNA (mRNA) molecules in specific regions of the DNA. During transcription, the nucleotide sequence of the DNA is "transcribed" into a corresponding nucleotide sequence of mRNA. 2. Translation: This stage occurs in the cytoplasm, within the ribosomes—the cellular organelles responsible for protein synthesis. Here, the nucleotide sequence of mRNA is translated into a sequence of amino acids that will make up the protein. The deciphering of the genetic code, namely the establishment of the correspondence between codons and amino acids, was accomplished by American biochemists such as M. Nirenberg and S. Ochoa.

genetic information — a set of characteristics that define an organism, inherited from its parents and stored in genetic structures (DNA, RNA) in the form of genetic code. This genetic code determines the organism's morphological structure, growth, development, metabolic processes, mental disposition, and susceptibility to diseases.

genetics — the science that studies the laws of heredity and variation in organisms, as well as the methods for manipulating these phenomena.

genome — the collection of genes contained within the haploid (single) set of chromosomes of a given organism.

genomics — a branch of genetics that aims to study the genomes of individual species at the molecular level, including their structure (structural genomics) and function (functional genomics). It also encompasses the application of genomic knowledge in areas such as genetic engineering, biotechnology, and gene therapy.

genotype — The genetic (hereditary) makeup of an organism, consisting of all its genes. In modern genetics, it is not considered a mere collection of independently functioning genes, but rather a unified system in which any gene can interact complexly with the others. All the properties and traits of an organism (its phenotype) are the result of the genotype being expressed under specific environmental conditions.

globule — (from the latin "Globulus" meaning "sphere") a colloidal particle with a spherical shape, or a macromolecule (such as a protein) that has been folded into a compact ball.

golgi apparatus — (named after C. Golgi) an intracellular network structure that plays a role in the production of certain cellular products (such as various secretions, collagen, glycogen, lipids, etc.) as well as in the synthesis of glycoproteins.

graph isomorphism — Graphs are called isomorphic if there exists a one-to-one correspondence between their vertices and edges, and moreover, if the corresponding edges connect the corresponding vertices.

guanine — a purine base that is found in the cells of all organisms as part of nucleic acids; it is one of the four "letters" that make up the genetic code.

H

homeostasis — an adaptive property of an organism (or system) that enables it to maintain certain indicators related to its functioning, even under changing external and internal conditions that may be critical (and even destructive to some aspects of the organism's structure or functions).

homologous sequences — two sequences that are similar due to a common evolutionary origin.

hydrophilicity — the ability of a substance or material to absorb water.

hydrophobicity — the property of a substance or material that prevents it from being wetted by water.

hydroxyl group — (OH) a –OH group that is part of the molecules of chemical compounds, such as water (H2O), alkalis (NaOH), and others.

I

implicative relationship — a triple vector whose components take values from the set {0, ½, 1}, where "½" represents a symbol of uncertainty (indeterminate truth value). Such a relationship is called implicative.

induction — (from latin inductio – meaning "to lead" or "to guide") is a logical process of deriving a hypothesis or a general statement from specific facts.

intron — A sequence that is not expressed (i.e., not translated into amino acids). This sequence does not encode any amino acids. It is a region of eukaryotic DNA (a gene) that is transcribed, but is subsequently removed from the primary transcript during splicing to form the functional mRNA. The sequences located on either side of the intron (exons) are then joined together.

intuition — A keen sense or perception, a subtle understanding that enables one to grasp the very essence of something.

L

law of large numbers — one of the principles of probability theory, which states that the combined effect of a large number of random factors will, under certain conditions, lead to a result that is almost independent of chance.

locus — (from Latin locus) The location of a specific gene on a chromosome or on a genetic map of the chromosome.

M

Markov chain — a special case of a Markov process. It consists of a sequence of random numbers, where the probabilities of each number occurring are related to one another; more precisely, the probability of a given value of a subsequent number depends only on the immediately preceding value, not on the full history of the sequence.

Markov process — an important special case of stochastic processes. A process is considered Markovian if the complete data required for making a prediction can be obtained by starting from the description of the process's current state; in other words, additional information about previous states does not contribute to the possibility of making a successful prediction.

messenger RNA — a transcript (a copy) of a gene that encodes a protein.

meaningful codons — 61 out of the 64 codons that encode 20 amino acids.

mitochondria — (from the Greek words "mitos" – thread, and "chondrion" – small particle/granule) are organelles found in both animal and plant cells. Oxidative-reductive reactions take place within mitochondria, providing energy for the cells. The number of mitochondria in a single cell can range from one to several thousand.

mitosis — (from the Greek word "Mitos" – thread) is a method of cell division that ensures an equal distribution of genetic material among daughter cells and the inheritance of chromosomes across successive generations of cells.

mutagenesis — the process by which hereditary changes, or mutations, occur in an organism. The basis of mutagenesis lies in alterations to the molecules of nucleic acids, which store and process genetic information.

mutagens — physical and chemical agents that cause hereditary changes, namely mutations: ionizing radiation and ultraviolet light, as well as various chemical compounds (such as mustard gas (sulfur mustard) and nitrous acid), among others.

mutations — (from the Latin mutatio – change, alteration) are natural or artificially induced changes in an organism's hereditary traits, resulting from alterations and disruptions in the genetic material of the organism – chromosomes and genes. Mutations are the basis for hereditary variation in living organisms.

N

nanometer — a unit of length equal to 10^{-9} meters, 10^{-3} micrometers, or 10 angstroms (Å).

native — (from latin "nativus" meaning innate) (biological) refers to something that is in its natural state, unmodified, and has retained the structure it possesses within a living cell (for example, a native protein).

negentropy — from the perspective of an "informational approach," is an antonymous concept derived from entropy; in other words, it is a notion that emphasizes the organizing and ordering aspects of a system, as opposed to disorder.

nucleic acids (polynucleotides) — high-molecular-weight organic compounds composed of nucleotide residues. Depending on the type of sugar present in their structure—deoxyribose or ribose—nucleic acids are classified as deoxyribonucleic acid (DNA) or ribonucleic acid (RNA), respectively.

nucleosome — the main structural unit of a chromosome. It consists of a histone core around which DNA is wrapped, with a length of 145 base pairs (bp), forming 1.75 turns around the histone core.

nucleotides — phosphoric esters of nucleosides; they consist of a nitrogenous base (purine or pyrimidine), a sugar (ribose or deoxyribose), and a phosphate group. Nucleotides are essential components of nucleic acids, coenzymes, and other biologically active compounds.

O

operon — a group of genes that are closely related to each other and together determine the synthesis of proteins, which are enzymes that act in successive stages of a particular biochemical process.

P

panspermia — (from "pan-" meaning all, and "sperma" meaning sperm) is the hypothesis that the seeds of living organisms are universally distributed throughout the universe, evolving over time. This idea has existed since ancient times until the early 20th century. Panspermia was also used to explain the emergence of life on Earth.

phenotype — The sum of all characteristics and traits of an organism that have been developed during its individual development. It is formed as a result of the interaction between the organism's genetic properties (genotype) and the environmental conditions in which it lives.

phyllotaxis — the distinctive, grid-like arrangement of leaves, seeds, petals, and scales in many plant species.

prokaryotes — (from the Latin "pro" – before, and the Greek "karyon" – nucleus) are organisms that, unlike eukaryotes, do not possess a well-defined cellular nucleus or a typical chromosomal structure. Genetic information is stored and transmitted through DNA.

proteins — high-molecular-weight organic substances composed of residues of 20 amino acids. They form the basis of all biological processes in living organisms. Proteins can be classified into simple types (proteins) and complex types (proteinoids), which contain non-protein components in addition to amino acids. In living organisms, proteins perform both structural roles (in building tissues and cellular components) and functional roles (such as acting as enzymes, hormones, pigments, etc.).

proteomics — the science that studies the set of proteins and their interactions within living organisms (a proteome is the collection of all proteins in an organism).

pseudogenes — non-functional, "silent" genes that arise as a result of mutations that disable normal, functional genes. (There are other, rarer mechanisms by which pseudogenes can form as well.).

purine bases — derivatives of the nitrogenous base purine (adenine and guanine). Their biological role in the life activities of all organisms is determined by their participation in the composition of nucleic acids.

pyrimidine bases — natural organic compounds derived from pyrimidine. Their biological significance in the life processes of all organisms lies in their role in the formation of nucleic acids, coenzymes, and other bioactive compounds.

R

reading frame — a nucleotide sequence that is expressed in codons, starting with a start codon and ending with a stop codon. Each DNA strand corresponds to three reading frames: one correct reading frame, which is also called the open reading frame (ORF), and two alternative (terminating) reading frames.

recognition — based on the comparison of features and characteristics of the object under study with those of other known objects. As a result, a conclusion is drawn regarding the most likely correspondence between them.

replication — (from the late Latin "replicatio" – meaning "repetition") is the process by which DNA molecules are duplicated with the participation of specialized enzymes. Replication ensures the accurate copying of the genetic information contained within DNA molecules and its transmission from one generation to the next.

ribonucleic acids (RNAs) — high-molecular-weight organic compounds that belong to the category of nucleic acids. They are composed of nucleotides, which include adenine, guanine, cytosine, and uracil, as well as the sugar ribose (in RNA, thymine is replaced by uracil; in DNA, ribose is replaced by deoxyribose). In cells of all living organisms, RNAs play a key role in the transmission and implementation of genetic information. There are three main types of RNAs: messenger RNAs (mRNAs), transfer RNAs (tRNAs), and ribosomal RNAs (rRNAs).

ribose — a monosaccharide that is present in all living cells as part of RNA.

ribosomes — intracellular particles composed of RNA and proteins; they are involved in the biosynthesis of proteins. They have been found in the cells of all living organisms.

S

site — a specific location within a DNA molecule.

somatic cells — the cells that make up the body of a multicellular organism, with the exception of reproductive cells (sex cells).

somatic mutations — non-hereditary mutations.

strain — a pure culture of microorganisms isolated from a specific source. Different strains of the same microorganism can vary in various properties (such as virulence or susceptibility to antibiotics).

synonymous codons — a group of triplets that encode the same amino acid.

T

recurrence — comes from the Latin "recursio", which means "to repeat or to return again".

termination codons — three nucleotide triplets that do not encode any amino acid. They signal the end of protein synthesis.

thymine — a pyrimidine base that is found in all living organisms as part of DNA. It is one of the four "letters" that make up the genetic code.

U

universality of the code — The code is universal if the same codon corresponds to the same amino acid in all organisms.

uracil — a pyrimidine base. It is found in all living organisms as part of uracil nucleotides and ribonucleic acids (RNA). It is one of the four "letters" of the genetic code; in DNA, thymine replaces uracil.

V

vacuoles — (from the Latin "vacuus" – meaning empty) are compartments found within animal and plant cells, as well as within single-celled organisms. There are two types of vacuoles: digestive vacuoles and contractile (pulsating) vacuoles. Contractile vacuoles regulate osmotic pressure and are responsible for removing waste products from the organism.

W

wobble rule (Degeneracy of the genetic code) — a rule in code theory: if two codons have the same first two nucleotides, and their third nucleotides belong to the same class (purine or pyrimidine), they encode the same amino acid.

Z

zygote — a cell formed by the fusion of gametes.

LIST OF REFERENCES

1. ["Roditel'skij Dom" (Parental Home) (newspaper), October 29, 2010, No. 24 (133)]. (in Russian).
2. [Under the Weight of the Genome]. NEWSWEEK. June 7–13, 2004. (in Russian).
3. *Gnatik E.N. [Human Genetics: The Past and the Future]. Moscow, LKI Publ., 2007. (in Russian).*
4. Sverdlov E.D. [Molecular Genetics Today: Great Achievements, Serious Problems, and Great Hopes]. Molekulyarnaya Genetika, Mikrobiologiya i Virusologiya [Molecular Genetics, Microbiology, and Virology]. 2013, No. 1. (in Russian).
5. *Lebedev S.A. [Induction as a Method of Scientific Knowledge]. Moscow, Izd-vo Moskovskogo Universiteta Publ., 1980. (in Russian).*
6. *Marks K., Engel's F. [Collected Works]. Vol. 20, p. 541. (in Russian).*
7. *Smirnov S.A. [Valuation of Intellectual Property]. Moscow, Finansy i Statistika Publ., 2003. (in Russian).*
8. [Populyarnaya Mekhanika (Popular Mechanics)]. September 2012, No. 9. (in Russian).
9. *Gupal V.M. [Methods for Recognition of Complex Systems: The Bayesian Procedure as the Optimal Recognition Procedure]. Moscow, 2005. (in Russian).*
10. *Gupal V.M. [Mathematical Methods for Analysis and Recognition of Genetic Information]. Moscow, RIOR; INFRA-M Publ., 2012. (in Russian).*
11. Krupskij M.A., Gupal V.M. [Diagnostics of Systems Using the Bayesian Recognition Procedure]. Kompetentnost' [Competence]. 2004, No. 4(16). (in Russian).
12. Gupal V.M., Krupskij M.A. [Bayesian Recognition Procedure as a Method for Diagnosing Technical Systems]. Pribory i Sistemy. Upravlenie, Kontrol', Diagnostika [Instruments and Systems: Management, Control, Diagnostics]. 2005, No. 3. (in Russian).
13. *Uspenskij V.A., Vereshhchagin N.K., Plisko V.E. [Introduction to Mathematical Logic]. Moscow, Fizmatlit Publ., 2002. (in Russian).*
14. *Dyugamel Zh.M.K. [Methods of Speculative Sciences]. Moscow, URSS Publ., 2010. (in Russian).*
15. *Krajzel' G. [Biography of Kurt Gödel]. Moscow–Izhevsk, Institut Komp'yuternykh Issledovanij Publ., 2003. (in Russian).*
16. *Podnieks K.M. [Around Gödel's Theorem]. Riga, 1981. (in Russian).*
17. *Biryukov B.V. [The Heat of Cold Numbers and the Passion of Impassive Logic]. Moscow, Znanie Publ., 1985. (in Russian).*
18. [Pavlus Dzh. Mashiny Neogranichennykh Vozmozhnostej. V Mire Nauki. 2012. No. 11]. (in Russian).
19. *Ge'ri M.R., Dzhonson D.S. [Computers and Intractability: A Guide to the Theory of NP-Completeness]. Moscow, Mir Publ., 1982. (in Russian).*
20. *Kozlov N.N. [Mathematical Analysis of the Genetic Code]. Moscow, Binom, Laboratoriya Znanij Publ., 2010. (in Russian).*
21. *Strygin V.Z. [Polynomial Algorithms for Solving Enumeration Problems]. Moscow State University (FVMiK): Avtoref. Diss. dokt. fiz.-mat. nauk. Moscow, 2000. (in Russian).*
22. Korolev L.N. [Evolutionary Computation, Neural Systems, Genetic Algorithms: Formal Problem Formulations]. Fundamental'naya i Prikladnaya Matematika [Fundamental and Applied Mathematics]. Vol. 15, Iss. 3. Moscow, 2009. (in Russian).
23. *Frolov I.T. [Philosophy and History of Genetics: Searches and Discussions]. Moscow, 2007. (in Russian).*
24. *Skirbekk G., Gil'e M. [History of Philosophy, Ch. 2: Sophists and Socrates]. Translated from English. Moscow, Vlados Publ., 2003. (in Russian).*
25. *Frolov I.T. [Essays on the Methodology of Biological Research]. Moscow, URSS Publ., 2012. (in Russian).*

26. Parmon V.N. [The Prebiotic Phase in the Origin of Life]. Vestnik RAN [Herald of the Russian Academy of Sciences]. 2002, No. 11, pp. 976–983. (in Russian).
27. *Stil E`., Lindli R., Blaunde`n R. [What If Lamarck Is Right? Immunogenetics and Evolution]. Moscow, Mir Publ., 2002. (in Russian).*
28. [Nauka i Zhizn' (Science and Life)]. 2010, No. 11, p. 40. (in Russian).
29. Lima-de-Fariya A. [In Praise of the "Stupidity" of the Chromosome: Confessions of a Rebellious Molecule]. Moscow, Binom, Laboratoriya Znanij Publ., 2012. (in Russian).
30. [Novyj Vek (New Century)]. October 2003, pp. 32–41. (in Russian).
31. *Nosov V.M. [What Future Awaits Humanity]. Pushkino, Graal' Publ., 1998. (in Russian).*
32. *Zarenkov N.A. [Biosymmetry]. Moscow, 2008. (in Russian).*
33. *E`ngel's F. [Anti-Dühring]. Moscow, 1957, p. 322. (in Russian).*
34. [Populyarnaya Medicinskaya E`nciklopediya (Popular Medical Encyclopedia)]. 1968, pp. 286–287. (in Russian).
35. Kordyum V.A. [What Is "Life" and What Can Follow from Our Conceptions of It]. Problemy Upravleniya i Informatiki [Problems of Control and Informatics]. 2012, No. 1. (in Russian).
36. *Vejsman M. [The Secret Life of the Body: The Cell and Its Hidden Possibilities]. Moscow, Vektor Publ., 2011. (in Russian).*
37. Kamenskij A.A., Maslova M.V., Graf A.V. [Hormones Rule the World]. Nauka i Mir [Science and the World]. 2010. (in Russian).
38. [Vestnik RAN (Herald of the Russian Academy of Sciences)]. Vol. 82, No. 6, 2012. (in Russian).
39. *Loveczkij G.I. [Philosophy and Mathematics: Higher Ideas and Numbers in the Ancient World and Antiquity]. Moscow, 2009. (in Russian).*
40. *Kastler G. [The Theory of Information in Biology]. Moscow, IL Publ., 1960. (in Russian).*
41. *Tyunyaev A.A. [Organizmika: The Fundamental Basis of All Sciences]. Moscow, In Publ., 2004. (in Russian).*
42. *Kastler G. [The Origin of Biological Organization]. Moscow, Mir Publ., 1967. (in Russian).*
43. *Buslenko N.P. [Modeling Complex Systems]. Moscow, Nauka Publ., 1978. (in Russian).*
44. *Ptashne M. [Gene Switch]. Moscow, Mir Publ., 1988. (in Russian).*
45. Kotov P. [The Main Question: What Is Most Important in the World]. Vse Yasno [All Is Clear]. August 2006, No. 8. (in Russian).
46. [Sovetskij E`nciklopedicheskij Slovar' (Soviet Encyclopedic Dictionary)]. Moscow, Sovetskaya E`nciklopediya Publ., 1982. (in Russian).
47. [Bol'shaya Rossijskaya E`nciklopediya (Great Russian Encyclopedia)]. Moscow, Bol'shaya Rossijskaya E`nciklopediya Publ., 2007. (in Russian).
48. *Polynin V.M. [A Prophet in His Own Land]. Moscow, Sovetskaya Rossiya Publ., 1969. (in Russian).*
49. *Kivi Berd. [The Theorem of Reverend Bayes]. 2001. (in Russian).*
50. Lindley D.V. Making Decisions. London: Wiley InterScience, 1971.
51. *Savchuk V.P. [Bayesian Methods of Statistical Estimation: Reliability of Technical Systems]. Moscow, Nauka Publ., 1989. (in Russian).*
52. *Nalimov V.V. [A Probabilistic Model of Language]. Moscow, Nauka Publ., 1974. (in Russian).*
53. *Nalimov V.V. [In Search of Other Meanings]. Moscow, Progress Publ., 1993. (in Russian).*
54. *Kremer N.Sh. [Probability Theory and Mathematical Statistics: A Textbook]. 3rd ed. Moscow, YUNITI Publ., 2007. (in Russian).*
55. Aleksandrovskaya L.N., Kuzneczov A.G. [The Bayesian Approach as a Set of Principles for Constructing Forms and Methods for Studying Objects. State

Standard USSR]. Partnery i Konkurenty [Partners and Competitors]. 2004, No. 1–2. (in Russian).
56. *Zel'ner A. [Bayesian Methods in Econometrics]. Moscow, Statistika Publ., 1980. (in Russian).*
57. *Zakrevskij A.D. [Logic of Recognition]. Moscow, URSS Publ., 2003. (in Russian).*
58. [Krasnaya Zvezda (Red Star)], No. 159, 164, August–September 2012. (in Russian).
59. *Brusenczov N.P. [The Use of Ternary Code and Three-Valued Logic in Digital Machines]. Moscow, MGU im. M.V. Lomonosova Publ., 1969. (in Russian).*
60. *Dzhonson Dzh.K. [Methods of Design]. Moscow, Mir Publ., 1986. (in Russian).*
61. *Bunge M. [Intuition and Science]. Moscow, Progress Publ., 1967. (in Russian).*
62. *Mikhalevich V.S. [Expert Assessments in Scientific and Technical Forecasting]. Kyiv, Naukova Dumka Publ., 1974. (in Russian).*
63. [Iskusstvennyj Intellekt (Artificial Intelligence): In 3 Volumes: Reference Book]. Ed. by D.A. Pospelov. Vol. 2: Modeli i Metody [Models and Methods]. Moscow, Radio i Svyaz' Publ., 1990. (in Russian).
64. Fedonin G.G., Rakhmaninova A.B., Korestelev Yu.D., Gel'fand M.S. [Study of DNA Binding by LacI Family Transcription Factors Using Machine Learning Methods]. Molekulyarnaya Biologiya [Molecular Biology]. 2011, Vol. 45, No. 4. (in Russian).
65. *Ivakhnenko A.G. [Long-Term Forecasting and Management of Complex Systems]. Kyiv, Tekhnika Publ., 1975. (in Russian).*
66. *Zozulya Yu.I. [Intelligent Neural Systems]. Moscow, Radio-Tekhnika Publ., 2003. (Series "Nejrokomp'yutery i ikh Primenenie" [Neurocomputers and Their Applications]). (in Russian).*
67. *Vernadskij V.I. [The Biosphere and the Noosphere]. Moscow, Ajris Press Publ., 2010. (in Russian).*
68. *Kant I. [Critique of Pure Reason]. Collected Works in 6 Volumes. Vol. 3. Moscow, Mysl' Publ., 1964. (in Russian).*
69. *Kant I. [Prolegomena to Any Future Metaphysics That Could Appear as a Science]. Collected Works in 6 Volumes. Vol. 4, Part 1. Moscow, Mysl' Publ., 1965. (in Russian).*
70. Charushnikov V.D. [Mathematization of Knowledge and Integration of Sciences]. Filosofiya v Sovremennom Mire: Dialog Mirovozzrenij [Philosophy in the Modern World: Dialogue of Worldviews]. Vol. 1. Rossijskij Filosofskij Kongress. N. Novgorod, 2012. (in Russian).
71. Moroz V.V. [On the Characteristics of Using Mathematical Models in Philosophical Knowledge]. Vol. 1. Rossijskij Filosofskij Kongress [Russian Philosophical Congress]. N. Novgorod, 2012. (in Russian).
72. Belostockaya S. [A Flash of Life]. Novyj Vek [New Century]. 2000. (in Russian).
73. *Ozhegov S.I. [Dictionary of the Russian Language]. Moscow, Sovetskaya E`nciklopediya Publ., 1975. (in Russian).*
74. *Jeffreys H. Theory of Probability. Oxford Clarendon, 1966.*
75. [Vse E`to Matematika (It's All Mathematics)]. V Mire Nauki [In the World of Science]. 2010, No. 5. (in Russian).
76. De Finetti B. Bayesianism: its role for both the foundations and applications of statistics. Internat. Statist. Rev. 1974. V.42, No.1. pp. 117–130.
77. Cornfield T. The Bayesian outlook and its application. Review Int. Stat. Inst. 1967. V. 35, No. 1. pp. 34–39.
78. Nejman D. [Two Breakthroughs in the Theory of Selecting Statistical Decisions]. Matematika. Periodicheskij Sbornik Perevodov Inostrannykh Statej [Mathematics: Periodical Collection of Translations of Foreign Articles]. Moscow, Mir Publ., 1964, Vol. 8, No. 2, pp. 113–140. (in Russian).
79. *Mizes Von R., Mathematical Theory of Probability and Statistics, New York: Acad. Press, 1964, 360 p.*
80. *Pupkov K.A., Kon'kov V.G. [Intelligent Systems]. Moscow, Izdatel'stvo MGTU im. N.E`. Baumana Publ., 2003. (in Russian).*

81. *Vitgenshtejn L. [Tractatus Logico-Philosophicus]. Moscow, IL Publ., 1995. (in Russian).*
82. Uajtkhed A. [Selected Works in Philosophy]. Translated from English. Moscow, Progress Publ., 1990. (in Russian).
83. *Markov A.A., Nagornyj N.M. [Theory of Algorithms]. Moscow, Nauka Publ., 1984, 432 pp. (in Russian).*
84. *Novikov P.S. [Constructive Mathematical Logic from a Classical Perspective]. Moscow, Nauka Publ., 1977. (in Russian).*
85. *Popper K. [Logic and the Growth of Scientific Knowledge]. Moscow, Progress Publ., 1983. (in Russian).*
86. Shestov L. [The Apotheosis of Groundlessness]. Izbrannye Sochineniya [Selected Works]. Moscow, Renessans Publ., 1993. (in Russian).
87. *Khill T.I. [Modern Theory of Knowledge]. Moscow, Progress Publ., 1965. (in Russian).*
88. *Klajn M. [Mathematics: The Loss of Certainty]. Moscow, Mir Publ., 1984. (in Russian).*
89. *Frolov I.T., Pastushnyj S.A. [Mendelism and the Philosophical Problems of Modern Genetics]. Moscow, LKI Publ., 2008. (in Russian).*
90. *Dubinin N., Gubarev V. [The Thread of Life]. Moscow, Atomizdat Publ., 1966. (in Russian).*
91. *Loewy A.G., Siekewitz P. (1970). Cell Structure and Function. Holl, Rinehart and Winston, London.*
92. *Romer A.S., Parsons T.S. (1978). The Vertebrate Body. W.B. Saunders, Philadelphia, USA.*
93. Schneider D. (1999). "Some levity in physics." Am. Sci. 87: 122–123.
94. *Kaminskaya E`.A. [General Genetics]. Minsk, 1982. (in Russian).*
95. *Popov V.V. [Genomics with Molecular-Genetic Foundations]. Moscow, URSS Publ., 2008. (in Russian).*
96. *Segal E., et al. (2006). A genomic code for nucleosome positioning. Nature 442: 772–778.*
97. Uillet E`. [Genetics Without Secrets: A Guide]. Moscow, 2008. (in Russian).
98. Laptev Yu.P. [Entertaining Genetics]. Moscow, Kolos Publ., 1982. (in Russian).
99. Tarantul V.Z. [The Human Genome: An Encyclopedia Written in Four Letters]. Moscow, 2003. (in Russian).
100. Nessa Ke`ri. [Epigenetics]. Feniks Publ., Rostov-on-Don, 2012. (in Russian).
101. Vilenkin N.Ya. [Combinatorics]. MCNMO Publ., 2013. (in Russian).
102. Ratner V.A. [Genetics, Molecular Cybernetics]. Lichnosti i Problemy [Personalities and Problems]. Novosibirsk, Nauka Publ., 2002. (in Russian).
103. Stent G., Ke`lindar R. [Molecular Genetics]. Moscow, Mir Publ., 1981. (in Russian).
104. Karasev V.A. [The Genetic Code: New Horizons]. St. Petersburg, Tessa Publ., 2003. (in Russian).
105. Me`tt Ridli. [Genome]. Moscow, E`ksmo Publ., 2010. (in Russian).
106. Frank-Kamenetzkij M.D. [The Most Important Molecule]. Moscow, 1983. (in Russian).
107. Nefedova L.N. [Application of Molecular Research Methods in Genetics: A Textbook]. Moscow, INFRA-M Publ., 2012. (in Russian).
108. [Khimiya i Zhizn' (Chemistry and Life)]. 1977, No. 4. (in Russian).
109. J.B.S. Haldane. A Banned Broadcast and Other Essays. London, 1946.
110. Morgan T. [The Structural Basis of Heredity]. Moscow–St. Petersburg, 1924, p. 10. (in Russian).
111. Pavlov I.P. [Complete Collected Works]. Vol. 3, Book One, pp. 124–125. (in Russian).
112. [Novaya Rossijskaya E`nciklopediya (New Russian Encyclopedia)]. Moscow, Infra-M Publ., 2008. (in Russian).

113. *Shcherbak V.I., 2003, Arithmetic inside the universal genetic code. BioSystems, Vol. 70, pp. 187–209.*
114. Gnedenko B.V. [A Course in Probability Theory]. Moscow, Nauka, Glavnaya Redakcziya Fiziko-Matematicheskoj Literatury Publ., 1988. (in Russian).
115. Kolmogorov A.N. [Probability Theory and Mathematical Statistics]. Moscow, Nauka Publ., 2005. (in Russian).
116. Gupal A.M., Vagis A.A. [The Inductive Approach in Mathematics]. Problemy Upravleniya i Informatiki [Problems of Control and Informatics]. 2002, No. 2. (in Russian).
117. Lesk A. [Introduction to Bioinformatics]. Moscow, Binom, Laboratoriya Znanij Publ., 2009. (in Russian).
118. Berezov T.T., Korovkin B.F. [Biological Chemistry]. (in Russian).
119. [Doklady Akademii Nauk RF (Reports of the Russian Academy of Sciences)]. Vol. 396, June 2004, No. 6, 740 pp. (in Russian).
120. [Nauka i Zhizn' (Science and Life)]. No. 6, 2012, p. 70. (in Russian).
121. Vol'kenshtejn M.V. [Crossroads of Science]. Moscow, Nauka Publ., 1972. (in Russian).
122. Garyaev P.P. [The Genetic Code Is More Complex Than Its Triplet Model]. Moscow, Obshhchestvennaya Pol'za Publ., 1997. (in Russian).
123. [V Mire Nauki (In the World of Science)]. No. 8, 2009. (in Russian).
124. Brandts D. [Structure and Stability of Biological Macromolecules]. Moscow, Mir Publ., 1973. (in Russian).
125. Fil'kenshtejn A.V. [Introduction to the Physics of Proteins: Lecture Notes]. 1999–2000. (in Russian).
126. Stakhov A.P. [Metaphysics and the Golden Ratio]. Akademiya Trinitarizma [Trinity Academy], Moscow, Issue No. 77-6567, April 10, 2006. (in Russian).
127. [Doklady Akademii Nauk RF (Reports of the Russian Academy of Sciences)]. 2011, Vol. 439, No. 1, pp. 30–32. (in Russian).
128. Markov A.A. [Probability Calculus]. Moscow, Gosizdat Publ., 1924. (in Russian).
129. Bernulli. [On the Law of Large Numbers]. Moscow, Nauka, Glavnaya Redakcziya Fiziko-Matematicheskoj Literatury Publ., 1986. (in Russian).
130. Feller V. [An Introduction to Probability Theory and Its Applications]. Vol. 1. Moscow, Mir Publ., 1984. (in Russian).
131. Kulagina O.S., Lyapunov A.A. [On the Problem of Modeling the Evolutionary Process]. Problemy Kibernetiki [Problems of Cybernetics]. Moscow, 1966. (in Russian).
132. Ivakhnenko A.G. [The Perceptron: A System for Pattern Recognition]. Kyiv, Naukova Dumka Publ., 1975. (in Russian).
133. Sergienko I.V., Beleczkij B.A., Vasil'ev S.V., Gupal A.M. [Prediction of the Secondary Structure of Proteins Based on Bayesian Recognition Procedures on Markov Chains]. Kibernetika i Sistemnyj Analiz [Cybernetics and System Analysis]. 2007. No. 2. (in Russian).
134. H. Fisher, Proc. Nat. Acad. Sci. US 51, 1285 (1964).
135. Gurov S.I. [Assessment of the Reliability of Classification Algorithms: A Textbook]. Moscow, MGU im. M.V. Lomonosova, Fakul'tet Vychislitel'noj Matematiki i Kibernetiki Publ., 2002. (in Russian).
136. Vapnik V.N. [Recovering Dependencies from Empirical Data]. Moscow, Nauka Publ., 1979. (in Russian).
137. Singer M., Berg P. [Genes and Genomes]. Vol. 1. Moscow, Mir Publ., 1998. (in Russian).
138. Vejr B. [Analysis of Genetic Data]. Moscow, Mir Publ., 1995. (in Russian).
139. Andrej Serdechnov. [Aging with a Clear Mind]. RBK Daily. No. 155, August 20, 2012. (in Russian).

140. Beleczkij B.A. [Modeling of Intracellular Distribution Processes Using Actively Charged Particles]. Problemy Upravleniya i Informatiki [Problems of Control and Informatics]. 2012, No. 3. (in Russian).
141. Mukhacheva T.A., Kovalev S.Yu. [Applied Bioinformatics]. UrFU. Yekaterinburg, 2012. (in Russian).
142. Vol'kenshtejn M.V. [Methods for the Study of Biopolymers]. Itogi Nauki i Tekhniki. Molekulyarnaya Biologiya [Advances in Science and Technology: Molecular Biology]. Moscow, 1979. (in Russian).
143. Sergienko I.V., Ryazanov V.V., Beleczkij B.A., Byc' A.V., Gupal A.M., Rzhepeczkij S.S. [Methods for Predicting the Spatial Structure of Proteins]. Kibernetika i Sistemnyj Analiz [Cybernetics and System Analysis]. 2010, No. 1. (in Russian).
144. Kuleshov A.Yu., Pisarzhevskij S.A. [Construction of a Large-Scale Model of Protein Structure]. Biofizika [Biophysics]. 2010. (in Russian).
145. Fil'kenshtejn A.V., Pticin O.B. [Protein Physics: A Course of Lectures]. 2002. (in Russian).
146. Karasev V.A., Luchinin V.V. [Introduction to the Design of Bionic Nanosystems]. Moscow, Fizmatlit Publ., 2009. (in Russian).
147. Sokolik V.V. [A Method for Modeling the Spatial Structure of a Protein Based on Its Determining Nucleotide Sequence]. Molekulyarnaya Biofizika [Molecular Biophysics]. 2010. (in Russian).
148. Spirin A.S. [Molecular Biology: The Structure of Ribosomes and Protein Biosynthesis]. Moscow, Vysshaya Shkola Publ., 1986. (in Russian).
149. Sergienko I.V., Gupal A.M. [Optimal Recognition Procedures and Their Application]. Kibernetika i Sistemnyj Analiz [Cybernetics and System Analysis]. 2007, No. 6. (in Russian).
150. Voevodin V.V. [Mathematical Foundations of Parallel Computing]. Moscow, Izd-vo MGU Publ., 1991. (in Russian).
151. Korobkov V.P. [Methods for Graph Partitioning into Minimally Connected Subgraphs and Their Use in Adaptive Information Processing]. Adaptacziya v Sistemakh Upravleniya [Adaptation in Computing Systems]. Riga, Zinatne, 1978. (in Russian).
152. Khinchin A.Ya. [R. Mises's Frequency Theory and Modern Ideas in Probability Theory]. Voprosy Filosofii [Questions of Philosophy]. 1961, Issue 1, pp. 91–102; Issue 2, pp. 77–89. (in Russian).
153. Prokhorov Yu.V. [Probability and Mathematical Statistics]. Moscow, BRE` Publ., 1999, 912 pp. (in Russian).
154. Ferguson, T.S.: Bayesian analysis of some nonparametric problems. Ann. Statist.—1973.—V.1. No. 2. pp. 209–230.
155. Dalat S.R. A note on adequacy of mixtures of Dirichlet processes. Sankhya, A., 1978. V. 40. No. 1. pp. 185–191.
156. Dalat S.R., Hall G.T. "On approximating parametric Bayes models". Ann. Statist. 1980. V 8., No. 5. pp. 664–672.
157. De Groot M. [Optimal Statistical Decisions]. Moscow, Mir Publ., 1974. (in Russian).
158. Kryanev A.V., Lukin G.V. [Mathematical Methods for Processing Uncertain Data]. Moscow, Fizmatlit Publ., 2003. (in Russian).
159. Boldin M.V., Simonova G.I., Tyurin Yu.N. [Sign Statistical Analysis of Linear Models]. Moscow, Nauka; Fizmatlit Publ., 1997. (in Russian).
160. E`l'yasberg P.E. [Measurement Information: How Much Is Needed? How to Process It?]. Moscow, Nauka Publ., 1983. (in Russian).
161. Kendal M., Styuart A. [Theory of Distributions]. Translated from English. Moscow, Nauka Publ., 1966. (in Russian).
162. [Veroyatnost' i Matematicheskaya Statistika. E`nciklopediya (Probability and Mathematical Statistics: An Encyclopedia)]. Moscow, 1999. (in Russian).

163. Gurov S.I. [Interval Estimation Based on the Principle of Consistency]. Moscow, Leninskie Gory, MGU, FVMiK Publ., 2006. (in Russian).
164. *Zaks Sh. [Theory of Statistical Inference]. Moscow, Mir Publ., 1975. (in Russian).*
165. Belen'kij V.Z., Zaslavskij A.A. [Foundations of the Theory of Fiducial Probabilities: The Principle of Inertia]. E`konomika i Matematicheskie Metody [Economics and Mathematical Methods]. Vol. 47, No. 3, 2011. (in Russian).
166. Beleczkij B.A., VagisA.A., Vasil'ev S.V., Gupal N.A. [Complexity of the Bayesian Inductive Inference Procedure: The Discrete Case]. Problemy Upravleniya i Informatiki [Problems of Control and Informatics]. 2006, No. 6. (in Russian).
167. Petukhov S.V. [Matrix Genetics, Algebras of the Genetic Code, Noise Immunity]. Moscow, 2008. (in Russian).
168. Bogatykh B.A. [Fractal Nature]. Moscow, URSS Publ., 2011. (in Russian).
169. Kopnin P. [Hypothesis and the Knowledge of Reality]. Kyiv, 1962. (in Russian).
170. Lobanov V.I. [Russian Logic vs. Classical Logic]. Kompaniya Sputnik Publ., Moscow, 2002. (in Russian).
171. Beleczkij A.Ya. [Combinatorics of Gray Codes]. Kyiv, 2003. (in Russian).
172. Archakov A.I. et al. [Study of the Mechanism of Macromolecule Function Based on Their Technological Applications in Biomedicine]. NII RAMN. 1983. (in Russian).
173. Holden C. "Alliance Launched to Model E.coli". Science, 2002. Vol. R. pp. 1459–1460.
174. Sverdlov E.D. [A View on Life Through the Window of the Genome]. Vol. 1. Moscow, Nauka Publ., 2009. (in Russian).
175. Stiven Kholl. [The Delayed Revolution]. V Mire Nauki [In the World of Science]. 2011, No. 1. (in Russian).
176. Uotson Dzh. [Molecular Biology of the Gene]. Moscow, Mir Publ., 1978. (in Russian).
177. Garyaev P.P. [Linguistic-Wave Genome: Theory and Practice]. Institut Kvantovoj Genetiki. Kyiv, 2009. (in Russian).
178. Romanovskij Yu.M., Tikhonov A.N. [Molecular Energy Converters in Living Cells: Proton ATP Synthase—A Rotating Molecular Motor]. Uspekhi Fizicheskikh Nauk [Physics Uspekhi]. 2010, No. 9, pp. 931–956. (in Russian).
179. Rubin A.B., Pytyeva N.F., Riznichenko G.Yu. [Kinetics of Biological Processes]. Moscow, Izd-vo Mosk. Universiteta Publ., 1977. (in Russian).
180. [Itogi (Summary)]. December 13, 2010. (in Russian).
181. Romanczev E. [Regular Miracles]. Moscow, Molodaya Gvardiya Publ., 1987. (in Russian).
182. Alberts B. et al. (1994). Molecular Biology of the Cell. Garland Publishing, N.Y., USA.
183. Efremova V.V., Aistova Yu.T. [Genetics]. Rostov-on-Don, Feniks Publ., 2010. (in Russian).
184. Vol'kenshtejn M.V. [Molecular Biophysics]. Moscow, Nauka Publ., 1975. (in Russian).
185. Ichas M. [The Biological Code]. Moscow, Mir Publ., 1971. (in Russian).
186. Patrushev L.I. [Gene Expression]. Moscow, Nauka Publ., 2000. (in Russian).
187. Dubinin N.P., Karpets I.I., Kudryavcev V.N. [Genetics, Behavior, Responsibility]. Moscow, 1982. (in Russian).
188. Samsonov F.A. [Fundamentals of Genetics in Defectology]. Moscow, 1980. (in Russian).
189. Baryshev Yu.I. [Heredity and Health]. Moscow, 1976. (in Russian).
190. Dmitrij Nazarov. [A Symphony of Life in Four Notes]. V Mire Nauki [In the World of Science]. No. 4, 2013, pp. 96–97. (in Russian).